INDUSTRY 4.0 - IMPACT ON INTELLIGENT LOGISTICS AND MANUFACTURING

Edited by **Tamás Bányai, Antonella Petrillo** and **Fabio De Felice**

Industry 4.0 - Impact on Intelligent Logistics and Manufacturing
http://dx.doi.org/10.5772/intechopen.76554
Edited by Tamás Bányai, Antonella Petrillo and Fabio De Felice

Contributors

Maria Pomffyova, Mária Rostašová, Vladimír Krajčík, Enrico Ferrera, Giuseppe Pacelli, Rosaria Rossini, Ilaria Bosi, Claudio Pastrone, Zongyan Wang, Andrés Montoya, Álvaro Guarín, Julián Mora, Salvador Cobos Guzman, Lucia Alonso Virgos, Maria Dolores Cima Cabal, Francisco Machio Regidor, Xue-Ming Yuan, John Runwei Cheng, Mitsuo Gen, Tamás Bányai

Notice

First published in London, United Kingdom, 2020 by IntechOpen
IntechOpen is the global imprint of INTECHOPEN LIMITED, registered in England and Wales, registration number: 11086078, 7th floor, 10 Lower Thames Street, London,
EC3R 6AF, United Kingdom
Printed in Croatia

British Library Cataloguing-in-Publication Data
A catalogue record for this book is available from the British Library

Additional hard and PDF copies can be obtained from orders@intechopen.com

Industry 4.0 - Impact on Intelligent Logistics and Manufacturing, Edited by Tamás Bányai, Antonella Petrillo and Fabio De Felice

Print: 978-1-83969-004-4
Online: 978-1-78985-453-4
eBook (PDF) ISBN: 978-1-78985-454-1

Meet the editors

Tamás Bányai received his Masters degree in 1993 and his PhD degree in 1999 from the University of Miskolc, where he is currently working as associate professor. He has more than 25 years of teaching and research experience in the design and control of materials handling systems and supply chain management, with special emphasis in heuristic optimization of large-scale systems. He has published over 150 research papers, book chapters, and conference proceedings. He has been a member and manager of more than 50 national and international research and development projects. Away from academia, his other interests include playing the piano and taking photographs.

Fabio De Felice is a professor at the University of Cassino and Southern Lazio, Italy, where he received his PhD in mechanical engineering. His current research focuses on multi-criteria decision-making analysis (with emphasis on AHP and ANP), digital manufacturing, and innovation management. He is the founder of the AHP Academy, which promotes the diffusion of the culture and methodologies of decision making. He is a member of the editorial boards of several international organizations and journals and he has authored/co-authored more than 100 articles in the areas of decision science and business management.

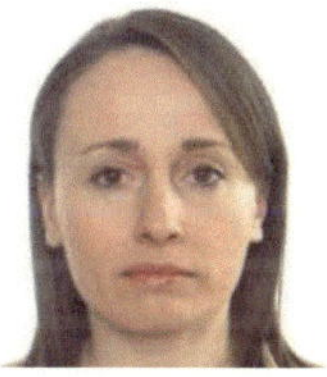

Antonella Petrillo is a professor at the Department of Engineering of the University of Naples "Parthenope", Italy. She received her PhD in mechanical engineering from the University of Cassino. Her research interests include sustainability, multi-criteria decision analysis, industrial plant, manufacturing, and safety. She serves as an international reviewer for over 50 international journals and she is a member of the Scientific Committee for PhD Programs, International Conferences and Proposal Evaluations.

Contents

Preface IX

Chapter 1 **Introductory Chapter: Industry 4.0 and Its Impact on Logistics - A Retrospective Review 1**
Tamás Bányai

Chapter 2 **Advantages of Learning Factories for Production Planning Based on Shop Floor Simulation: A Step towards Smart Factories in Industry 4.0 11**
Montoya Andrés, Guarín Álvaro and Mora Julián

Chapter 3 **Impact of Industry 4.0 on Inventory Systems and Optimization 27**
Xue-Ming Yuan

Chapter 4 **Cyber-Physical System Architecture for Minimizing the Possibility of Producing Bad Products in a Manufacturing System 39**
Salvador Cobos Guzman, Maria Dolores Cima Cabal, Francisco Machio Regidor and Lucia Alonso Virgos

Chapter 5 **Leveraging Internet-of-Things to Support Circular Economy Paradigm in Manufacturing Industry 53**
Giuseppe Pacelli, Enrico Ferrera, Rosaria Rossini, Ilaria Bosi and Claudio Pastrone

Chapter 6 **Parallel Genetic Algorithms with GPU Computing 69**
John Runwei Cheng and Mitsuo Gen

Chapter 7 **Digital Twin Technology 95**
Zongyan Wang

Chapter 8 **The Role of Spin-Off Companies in the Technology Transfer and IS Management Potential in Developing a Sharing Economy 115**
Mária Pomffyová, Mária Rostašová and Vladimír Krajčík

Preface

This book offers a selection of chapters in the field of the fourth industrial revolution, promoting new research results in the field, building on other publications in the literature. Authors from ten countries (China, Colombia, Czech Republic, Hungary, Italy, Japan, Singapore, Slovakia, Spain, and USA) have published worked examples and case studies from their research in the field. Implementing Industry 4.0 technologies and solutions is a strategic objective of manufacturing and service companies to foster competitiveness on the global market. The Industry 4.0 as an industrial ecosystem is a potential glue that can hold interconnected logistics systems and value making chains together. In a brief description, the book has eight topics, determined by the application domain and Industry 4.0 technologies and impacts:

Industry 4.0 and its impact on logistics: The globalization of economy led to an increased transportation and material handling intensity that is perceivable in purchasing, production, distribution, and recycling. Within the frame of this chapter, the main research directions of Industry 4.0 are shown focusing on logistic applications.

Advantages of Learning Factories for Production Planning based on shop floor simulation: In an Industry 4.0 industrial ecosystem, production planning and control strategies will be performed by humans backed by computational tools for decision making. In this chapter, shop floor simulation is identified as a tool for planning and controlling production and the trends in the construction of the Learning Factories are shown. This work also proposes the realization of a digital model in EAFIT University Learning Factory towards digital learning factory.

Impact of Industry 4.0 on Inventory Systems and Optimization: Industry 4.0 opens up new horizons for inventory optimization and pushes the boundaries, especially in the automotive industry. Within the frame of this chapter, the possible changes to factories, products, customers, and businesses in the environment of Industry 4.0 are discussed. The new integrative research and development framework for inventory systems and optimization is proposed.

Cyber-physical System Architecture for minimizing the possibility of producing bad products in a manufacturing system: The implementation of cyber-physical systems in an Industry 4.0 environment can increase the level of productivity in a manufacturing system. This chapter proposes a possible architecture of a generic manufacturing system including agile production, lean manufacturing, and statistical approaches.

Leveraging Internet-of-Things to support Circular Economy paradigm: New models based on Industry 4.0 technologies support the Circular Economy paradigm, through the development of a marketplace for connecting buyers and sellers of manufacturing services, raw materials, and products towards building global supply chains. This approach is explained within the frame of this chapter.

Parallel Genetic Algorithms with GPU Computing: Parallelizing heuristic and metaheuristic algorithm is a key factor of big data problems in the Industry 4.0 environment. In this chapter, the authors give a concise overview of selective works of Parallel GAs on GPU from the perspective of GPU architecture. Many optimization problems of cyber-physical systems will benefit from the described parallel GAs accelerated by GPU computing.

Digital Twin Technology: Digital twin technology is the core technology of realizing Cyber-Physical Systems. This chapter introduces the generation and characteristics of digital twin technologies, the relationship between Digital Twin and Digital Thread, and the implementation of the product digital twin model.

The role of spin-off companies in the technology transfer and IS management potential in developing a sharing economy: Based on best practice results, the chapter recommends practices and recommendations that should eliminate barriers to successful spin-offs and start-ups as well as SMEs doing business in competitive practice. Based on the synthesis of research outputs, the recommendations for technology transfer and regulatory treatment of sharing economy are summarized.

The aim of this book is to help students at the level of BSc, MSc, and PhD as well as managers and researchers, to understand and appreciate the concept, design, and implementation of Industry 4.0 solutions. Based on these considerations, the book may have a large impact in the scientific community.

The editors thank the research authors of the chapters for their scientific contribution. The chapters were edited and published following a rigorous selection process. Also, it is a pleasure thanking and acknowledging the help of many individuals from the editorial process that made this book possible.

Tamás Bányai
University of Miskolc
Department of Materials Handling and Logistics
Hungary

Antonella Petrillo
University of Napoli "Parthenope"
Department of Engineering
Italy

Fabio De Felice
University of Cassino and Southern Lazio
Department of Civil and Mechanical Engineering
Italy

Introductory Chapter: Industry 4.0 and Its Impact on Logistics - A Retrospective Review

Tamás Bányai

Additional information is available at the end of the chapter

http://dx.doi.org/10.5772/intechopen.89387

1. Introduction

The increased complexity of customers' demands led to the globalisation of economy. This caused increased transportation and material handling intensity which is perceivable in all four functional parts of logistics: purchasing, production, distribution and inverse processes. The globalisation, the demographical changes, the urbanisation and the sustainability influence the expected performance of manufacturing and service processes. The Industry 4.0 and its technologies offer new solutions to fulfil customers' demands, while economical, ecological, social and engineering impacts are taken into consideration.

Within the frame of this short introductory chapter, the main research directions of Industry 4.0 are shown focusing on logistic applications. The following questions are answered with a short literature review: Who is doing what? Who first did it or published it? What are research gaps?

2. Conceptual framework and review methodology

Within the frame of this retrospective literature review, the systematic literature review (SLR) methodology of Cronin et al. [1] was used including the following aspects: define research questions, select sources from Web of Knowledge, reduce the number of articles, identify the main research directions, define a methodology to analyse the chosen articles, discuss the main scientific results and identify the scientific gaps and bottlenecks.

The relevant terms were defined. The first search using the topic 'Industry 4.0' keyword resulted to 1066 articles, but a wide number of the articles have no relation with logistics;

therefore, the following keywords were used to search in the Web of Knowledge database: the topic 'Industry 4.0' and topic 'logistics'. Initially, 59 articles were identified. The search was conducted in July 2019; therefore, new articles may have been published since then.

3. Descriptive analysis

The articles can be classified depending on the research area. **Figure 1** shows the classification of these 59 articles considering 10 subject areas. This classification shows the majority of engineering and management sciences, while the operational research and computer sciences define the importance of computational methods related to the design of logistics and supply chain solutions in Industry 4.0 environment. The classification shows that sustainability, energy efficiency and environmental impact of logistic solutions of cyber-physical systems get more and more importance.

As **Figure 2** demonstrates, Industry 4.0 and its effect on logistics and supply chain solutions have been researched in the past 5 years. The first article in this field was published in 2015 in the field of cloud computing [2], and it was focusing on the importance of cloud-assisted solutions of the emerging Industry 4.0 theory focusing on logistics and sales. The number of published papers has been increased; it shows the importance of this research field, especially in automotive industry.

As **Figure 3** demonstrates, most of the articles were published in journals with sustainability and manufacturing topics. The distribution of journals shows that the research of logistics and supply chain solutions in cyber-physical systems and Industry 4.0 environment is a multidisciplinary topic.

In the following step, the 59 articles were reduced after reading them. Articles whose topic cannot address the effect of Industry 4.0 on logistics were excluded, and some

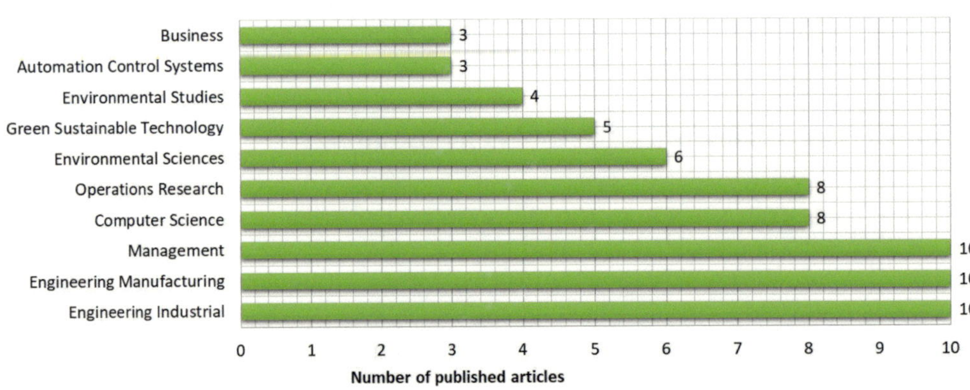

Figure 1. Classification of articles considering subject areas based on search in Web of Knowledge database using the topics 'Industry 4.0' and 'logistics'.

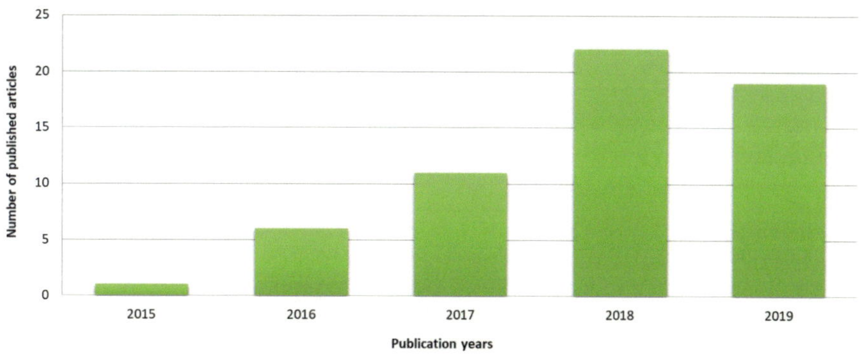

Figure 2. Classification of articles by year of publication based on search in the Web of Knowledge.

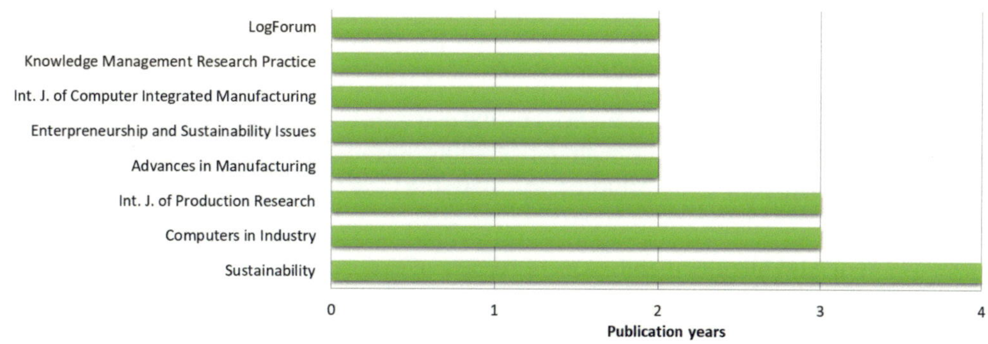

Figure 3. Distribution of Industry 4.0 and logistic-related articles in journals, based on search in the Web of Knowledge.

articles were added from different research platforms. After this reduction and addition, we got 18 articles. The next section describes the content analysis of these articles and defines a framework of available sources related to the impact of Industry 4.0 in the field of logistics.

4. Content analysis

The fourth industrial revolution is also known as 'Industry 4.0', 'Smart Manufacturing' or 'Integrated Industry', but all of these names cover the integration potential of value making chains from design, through production, to distributions and inverse processes. This potential is especially focused in just-in-time and just-in-sequence processes of automotive industry [3]. Industry 4.0 has a great impact on logistics of production [4] and service processes [5]. The lean paradigm (or lean manufacturing) is linked to the development and operation of cyber-physical systems, because it is widely regarded as an efficient improvement tool in Industry 4.0 environment [6]. Another aspect of the importance of lean manufacturing in

Industry 4.0 is that both cyber-physical systems and lean manufacturing can be characterized with decentralized control [7]. The globalisation of the economy led to the increased importance of sustainability and the minimization of environmental impact; therefore Industry 4.0 should include not only purchasing, manufacturing and distribution processes but also the inverse operations of the closed-loop economy [8]. The realisation of cyber-physical systems requires the integration of networking, organisational development, structural frame conditions and sustainability which in fact has a great impact on the logistic solutions in manufacturing and services [9]. The first-mile and last-mile operations of a supply chain solution represent complex optimization problems, because routing, assignment and scheduling problems should be taken into consideration. The Industry 4.0 technologies make it possible to make real-time decisions in the first-mile and last-mile processes in order to design a cost-efficient sustainable collection or distribution process. These design problems are NP-hard; therefore heuristic and metaheuristic solutions are required to solve them [10, 11]. Another revolutionary design method is the online optimization. Online optimization can support to solve real-time challenges [12] in complex supply chain processes or in-plant material handling systems.

Logistics and value chain will be influenced by the new solutions of Industry 4.0 technologies. Researches based on Porter's value chain model show that an increased economic sustainability can be achieved with the application of IoT tools, Big Data technologies and cyber-physical systems [13].

Bressanelli et al. [14] identified eight functionalities of Industry 4.0 technologies (improving product design, attracting target customers, monitoring and tracking product activity, providing technical support, providing preventive and predictive maintenance, optimising the product usage, upgrading the product, enhancing renovation and end-of-life activities) and analysed their impact on closed-loop value drivers, like resource efficiency, lifespan and closing the loop. These eight functionalities of Industry 4.0 technologies are important not only for manufacturing but also for logistic operations and supply chain processes. Case studies validate the impact of Industry 4.0 technologies on supply chain solutions. These case studies are based on manufacturing processes of Logistics Knapp AG, Nova Chemicals, BMW, Stratasys and Bosch [15].

The logistic operations include not only the in-plant supply processes but also the material handling operations of warehousing. The increased complexity of manufacturing processes led to increased complexity of component portfolios which led to complex warehousing processes. In the case of real-time data and contextual information, these complex warehouse processes can be improved through Industry 4.0 technologies [16]. The success of applied Industry 4.0 technologies depends on the characteristics of manufacturing environment [17]. The manufacturing environment can be described with the level of automation, level of product repetitiveness or type of products.

Researches show that the most important Industry 4.0 technologies applied in the field of logistics are focusing on the support of product identification and increase of traceability of products and resources [18]. **Table 1** shows the research framework represented by the reviewed articles from objective, methodology and case studies point of view.

	Cloud computing	Supply chain	Lean manufacturing	Closed-loop economy	Organisation	Heuristics	Online optimization	Real-time decision	Internet of Things	Optimization	Service	Manufacturing	Case studies
Yue et al. (2015) Cloud-assisted industrial cyber-physical systems [2]	X												
Hofmann and Rusch (2017) Industry 4.0 and the current status as well … [3]		X										X	
Sanders et al. (2016) Industry 4.0 implies lean manufacturing … [6]			X									X	
Jabbour et al. (2018) Industry 4.0 and the circular economy [8]		X		X									
Prause and Atari (2017) On sustainable production networks for … [9]		X			X							X	
Banyai (2018) Real-time decision making in first mile and … [10]		X				X		X		X	X		

	Cloud computing	Supply chain	Lean manufacturing	Closed-loop economy	Organisation	Heuristics	Online optimization	Real-time decision	Internet of Things	Optimization	Service	Manufacturing	Case studies
Nagy et al. (2018) The role and impact of Industry 4.0 and the … [13]		X			X						X	X	
Buer et al. (2018) The link between Industry 4.0 and lean … [7]			X									X	
Bressanelli et al. (2018) Exploring how usage-focused business … [14]				X	X								
Banyai et al. (2018) Smart scheduling: an integrated first mile … [11]		X				X		X		X	X		
Szozda (2017) Industry 4.0 and its impact on the functioning … [15]		X			X						X		X
Lee et al. (2018) Design and application of Internet of … [16]						X			X		X	X	

	Cloud computing	Supply chain	Lean manufacturing	Closed-loop economy	Organisation	Heuristics	Online optimization	Real-time decision	Internet of Things	Optimization	Service	Manufacturing	Case studies
Strandhagen et al. (2017) The fit of Industry 4.0 applications … [17]									X			X	X
Trappey et al. (2017) IoT patent roadmap for smart logistic … [18]									X		X		
Dunke et al. (2018) Time traps in supply chains … [12]						X	X			X			
Tamás et al. (2017) New challenges for quality assurance … [5]		X	X								X		
Nagy et al. (2018) Impact of Industry 4.0 on production logistics [4]		X										X	

Table 1. Research framework and main topics of logistic-related research in Industry 4.0.

5. Consequences of literature review

The above described short literature review shows that the application of Industry 4.0 technologies offers innovative solutions for both manufacturing and logistic processes. Cloud and fog computing supports the solution of Big Data problems. Heuristic, metaheuristic and online optimization make it possible to solve NP-hard problems of complex logistic processes and supply chains. The adoption of bar codes and RFID technologies focuses on identification and traceability of products, resources and processes. In logistic processes, the importance of real-time decisions gets more and more importance, because real-time decisions can increase the efficiency of resources through an improved flexibility. Why is it necessary to put so much effort into the application of Industry 4.0 technologies? However, the application of Industry 4.0 technologies in logistics leads to new manufacturing technologies and logistic solutions, but the design and operation of these new processes and systems required sophisticated models and methods to build complex, efficient, sustainable solutions.

Author details

Tamás Bányai

Address all correspondence to: alttamas@uni-miskolc.hu

University of Miskolc, Hungary

References

[1] Cronin P, Ryan F, Coughlan M. Undertaking a literature review: A step-by-step approach. British Journal of Nursing. 2008;**17**(1):38-43

[2] Yue XJ, Cai H, Yan HH, Zou CF, Zhou KL. Cloud assisted industrial cyber-physical systems: An insight. Microprocessors and Microsystems. 2015;**39**(8):1262-1270

[3] Hofmann E, Rusch M. Industry 4.0 and the current status as well as future prospects on logistics. Computers in Industry. 2017;**89**:23-34

[4] Nagy G, Illés B, Bányai Á. Impact of Industry 4.0 on production logistics. IOP Conference Series: Materials Science and Engineering. 2018;**448**:012013

[5] Tamás P, Illés B, Dobos P, Skapinyecz R. New challenges for quality assurance of manufacturing processes in Industry 4.0. Solid State Phenomena. 2017;**261**:481-486

[6] Sanders A, Elangeswaran C, Wulfsberg J. Industry 4.0 implies lean manufacturing: Research activities in Industry 4.0. Function as enablers for lean manufacturing. Journal of Industrial Engineering and Management. 2016;**9**(3):811-833

[7] Buer SV, Strandhagen JO, Chan FTS. The link between Industry 4.0 and lean manufacturing: Mapping current research and establishing a research agenda. International Journal of Production Research. 2018;**56**(8):2924-2940

[8] Jabbour ABLD, Jabbour CJC, Godinho M, Roubaud D. Industry 4.0 and the circular economy: A proposed research agenda and original roadmap for sustainable operations. Annals of Operations Research. 2018;**270**(1-2):273-286

[9] Prause G, Atari S. On sustainable production networks for Industry 4.0. Entrepreneurship and Sustainability Issues. 2017;**4**(4):421-431

[10] Banyai T. Real-time decision making in first mile and last mile logistics: How smart scheduling affects energy efficiency of hyperconnected supply chain solutions. Energies. 2018;**11**(7):1833

[11] Banyai T, Illes B, Banyai A. Smart scheduling: An integrated first mile and last mile supply approach. Complexity. 2018:5180156

[12] Dunke F, Heckmann I, Nickel S, Saldanha-da-Gama F. Time traps in supply chains: Is optimal still good enough? European Journal of Operational Research. 2018;**264**(3):813-829

[13] Nagy J, Olah J, Erdei E, Mate D, Popp J. The role and impact of Industry 4.0 and the Internet of things on the business strategy of the value chain—The case of Hungary. Sustainability. 2018;**10**(10):3491

[14] Bressanelli G, Adrodegari F, Perona M, Saccani N. Exploring how usage-focused business models enable circular economy through digital technologies. Sustainability. 2018;**10**(3):639

[15] Szozda N. Industry 4.0 and its impact on the functioning of supply chains. Log Forum. 2017;**13**(4):401-414

[16] Lee CKM, Lv YQ, Ng KKH, Ho W, Choy KL. Design and application of Internet of things-based warehouse management system for smart logistics. International Journal of Production Research. 2018;**56**(8):2753-2768

[17] Strandhagen JW, Alfnes E, Strandhagen JO, Vallandingham LR. The fit of Industry 4.0 applications in manufacturing logistics: A multiple case study. Advances in Manufacturing. 2017;**5**(4):344-358

[18] Trappey AJC, Trappey CV, Fan CY, Hsu APT, Li XK, Lee IJY. IoT patent roadmap for smart logistic service provision in the context of Industry 4.0. Journal of the Chinese Institute of Engineers. 2017;**40**(7):593-602

Chapter 2

Advantages of Learning Factories for Production Planning Based on Shop Floor Simulation: A Step towards Smart Factories in Industry 4.0

Montoya Andrés, Guarín Álvaro and Mora Julián

Additional information is available at the end of the chapter

http://dx.doi.org/10.5772/intechopen.79930

Abstract

In future industry, defined as Industry 4.0, production planning and control strategies will be executed by human beings backed by computational tools for decision making; One of these tools is shop floor simulation, and a natural scenario to learn about how to use it for productive processes design and control are the Learning Factories. In this chapter, shop floor simulation is identified as a tool for planning and controlling production, also a state of the art about its implementation is exposed in academic and industrial environments. In addition, the trends in the construction of the Learning Factories are shown, and some aspects about how they can be used for shop floor simulation. This work also proposes the realization of a digital model in EAFIT University Learning Factory as a first step towards digital learning factory.

Keywords: digital learning factory, shop floor simulation, plant planning and controlling, discrete event simulation, Industry 4.0

1. Introduction: Learning Factories as simulated environments and innovation transfer actors

When a production system is being designed, builded or even working, there are several behaviors that cannot be described by exact mathematical equations (analytical models); for this reason, plant designers commonly use Virtual Design (VD), Shop Floor Simulation (SFS), heuristics or metaheuristic methods to predict possible critical situations once system is perturbed by use of supply chain, machinery, workstations cells, buffering and response time, equipment capabilities, etc. Some heuristic approaches are, for example, CRAFT algorithm used by Prasad et al. to optimize plant layout cost by machinery replacement [1] and Black Hole Algorithm used by Veres et al. to find optimal supply chains structure [2], which provide possible solutions in complex systems. However, SFS also allows to evaluate several system responses and to determinate optimal working conditions [3].

Simulation refers to Discrete Event Simulation (DES) which describe a model performance. According to Shanon [4], simulation is to design a real system model and carry out some experiences to learn its behavior and to evaluate strategies to understand how it works; it means that, in order to learn about system behavior, it is necessary an environment characterized by a high degree of fidelity regarding real system. That characteristic is typical of action-oriented learning methodologies such as role play [5], virtual reality [6], simulation [7] and Learning Factories (LF). Additionally, simulation has recently become a best valued teaching strategy by students [8], and LF are effective simulated environment tools for skills development through experiential learning [9]; its safe environments allow to face real industrial challenges with minimal abstractions [10] to replay to real situations in different knowledge fields during training process.

On the other hand, companies call for latest technical and technological knowledge and LF offer high potential for innovation transfer and application-oriented innovation platform of product and production processes research, and subsequent transmission of these innovations [11].

This article aims to provide a general state of the art of SFS based production planning in LF, seeks to define medium term research needs and proposes a research project to develop the EAFIT University LF digital model. It will be the starting point in the way to convert this LF in a digital Smart Factory towards Industry 4.0.

2. Shop floor simulation based production planning and control in Industry 4.0

SFS is commonly used to get information about production systems and is based on discrete events information related to resources flow and availability in the plant. In this context, a resource is everything needed for production (supplies and raw materials) and plant are the facilities used to transform those resources into a finished product (or service). Production system also involves other company areas not involved directly in product transformation but without whose existence production process would be much more complex and inefficient, for example: supply warehouse, packing, shipping, maintenance, etc. Interactions between all these areas are interesting for decision making and SFS lets to evaluate production capacity, inventory control, logistics, etc.

Currently, SFS role in Industry 4.0 context includes "real time" production programming and control. Some authors have suggested that using production system latest information and making simulations to predict events in the future it is possible to improve system performance by dynamically adjusting programming policies and control strategies [12, 13], it places SFS as an essential element in one of the four Industry 4.0 components, namely "smart factories" [14].

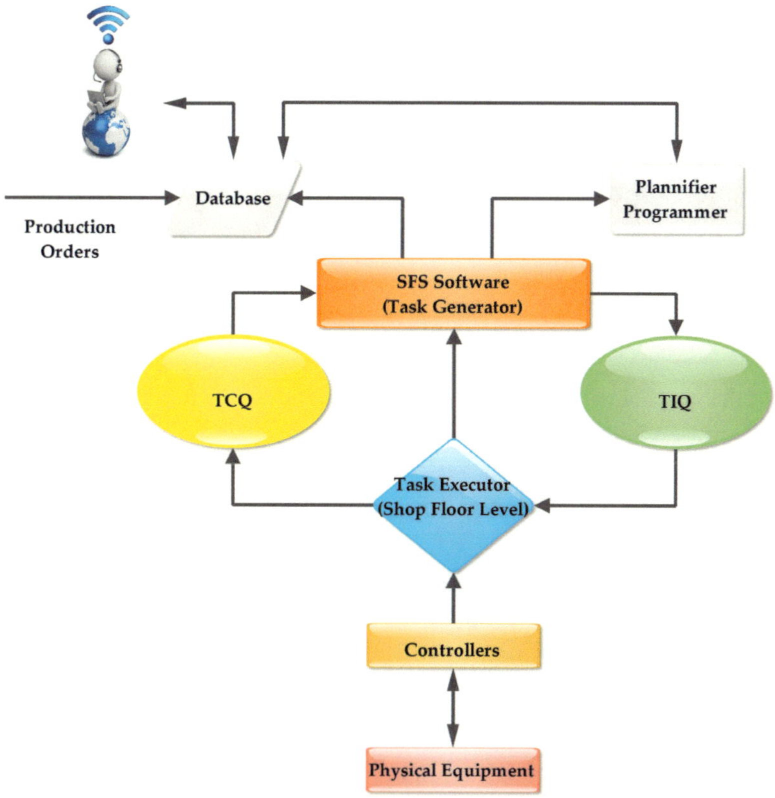

Figure 1. SFS based production control architecture in Industry 4.0. ([16] p. 382, 2005).

Figure 1 shows an example of SFS based production control architecture [15]. There is a task generation software used to develop simulation model which obtain production master programs (for example parts orders) and process plans from a database. The database keeps track part orders and quantity. Simulation control manufacturing system sends and receives messages and uses communication links to task executor; it performs execution functions at floor level and monitor part status in the system. Executor receives instructions (messages) from simulation and, according to system status, sends messages to physical equipment controllers. Once a task message has been sent, both executor and simulation wait for a "completed ok" message from controller. When executor receives "completed ok" message, it sends a similar message to simulation, and simulation knows that current task was completed. Simulator and execution modules communicate through a Task Initiation Queue (TIQ) and a Task Completion Queue (TCQ). Simulation uses TIQ to order executor perform specific tasks and receives completion messages through the TCQ. Son et al., [16] and Rao et al. [17] shows others SFS software use for controlling production process.

Digital models used in SFS are useful not only to control but also to design production plants layout. These models are usually stochastic-nature statistical distributions and are used to perform design variables analysis, control strategies and system performance estimation using commercial computer programs, for example: Arena™, AutoMod™, ProModel™, Tecnomatix Plant Simulation™, FlexSim™, Witness™, AnyLogic™, Process Model™, etc. Recently, these tools have been classified according to their popularity and tools [18].

Other interesting subject in Industry 4.0 is flexible production plants, it means "systems that allow changing product without high impact on costs either by sequence change programming" [19]. This flexibility can be understood as production lines flexibility or production processes flexibility. In general terms, if a good resources availability is achieved it is possible to maximize production process; This can be achieved adding new technology or using plant redistribution, and here SFS takes special interest because it let more easy decision making and it is much easier to achieve in a flexible LF where it is possible to reconfigure plant distribution. This is one of the greatest advantage in a LF for future research.

3. Digital Learning Factories implementation trends

A digital LF is an integrated IT environment where all real LF resources, processes and products are tracked on a digital model. Otherwise, virtual LF provide visual software tools trough Virtual Reality (VR) or Augmented Reality (AR) technology [20] to improve Digital LFs. Some LFs have implemented virtual games to evaluate students learning after training [21]. With these technologies, it is possible to carry out digital simulation, simulate tasks or evaluate alternative designs before production start [22].

LFs have become widespread in recent years, particularly in Europe [23], and have adopted many installation ways varying in size, scope, function and sophistication. However, all of they have main objective to improve students and industrial users learning experience. Abele et al. [24] present a right sounding classifying LFs worldwide according to:

Learning Factories for production process enhancement: They deal with lean methods and principles, such as just-in-time, line balancing, problem solving or work optimization.

Learning Factories for reconfiguration, production and plant layout design: They deal with reconfigurability related to plant layout design.

Learning Factories for energy and resources use efficiency: They deal with the relationship between energy consumption, resources and production.

Learning Factories for Industry 4.0: They focus on production digitization and Internet of Things (IoT) researching and technology transfer.

Learning Factory applied concept: They deal with knowledge triangle vertices integration: education, research and innovation.

Learning Factories for other purposes: For logistic optimization. For management and organization, automation technologies and sustainability.

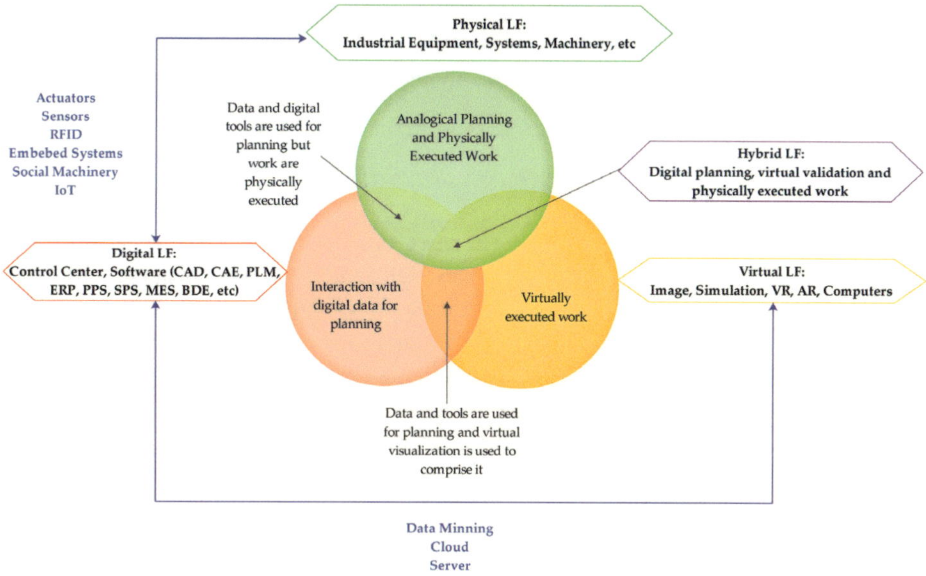

Figure 2. Physical, digital and virtual LF. ([24] p. 816).

This literature review shows there is an important trend towards facilities and production processes digitalization. Digital [25] and virtual LFs have been developed [26], even equipped with SFS software for material flow, tail and bottleneck analysis [27].

Figure 2 interrelation between infrastructure and physical interface in digital and virtual LFs. In this context, the SFS brings LF infrastructure to digital level.

First Colombian LF was inaugurated last September 27th, 2017. This LF is located at EAFIT University in Medellín City, and its main goal is to perform experiential learning practices with university and industry students to enhance their skills facing up new local production processes challenges, and help to transform conventional companies into smart companies [28] This LF is equipped with SFS such as Tecnomatix Plant Simulation and FlexSim for production processes improvement. Thus, next step towards Industry 4.0 once this LF was inaugurated is SFS implementation in its own Digital Model.

4. How to use Learning Factories in plant layout simulation?

Plant layout simulation difficult due to the following aspects:

1. High time consumption in digital modeling: Large time amount must be invested in the digital modeling.

2. Digital modeling validation: Sometimes it is necessary to build expensive pilot plants or to carry out experiments in a section or all production line causing delays or stoppages. This is due to flexibility lack in the real plant to change its distribution.

3. Planning and experimenting: Experiments sometimes requires redesign experiment itself to input or output not considered variables, which in many cases involves rethinking production strategies.

Thus, a LF is a right scenario to learn how to carry out plant layout planning based on SFS because it solves mentioned drawbacks: test model, validation and experimentation. This subject has been studied in digital LFs as noted in Section 2, however, although several digital LFs are equipped with SFS software, it is not common to find publications showing SFS use in LF layout planning.

So far, there is a worldwide trend to immerse plant layout designers on real, digital or virtual simulated environments to help them decision making. However, even for experienced staff, simulation can be a very long process, and transmitting that knowledge to undergraduate, postgraduate students or even business people usually takes a long time, mainly due to people focus more on simulation model construction than system behavior understanding.

In an LF, plant layout simulation can be carried out following methods described by Banks [29] and Tako [30]; their contributions, contextualized to plant layout are summarized in **Figure 3**. They suggest contemplating six phases to reproduce discrete events during productive system operation; therefore, a student, instructor or professional (henceforth called analyst) should focus his attention on:

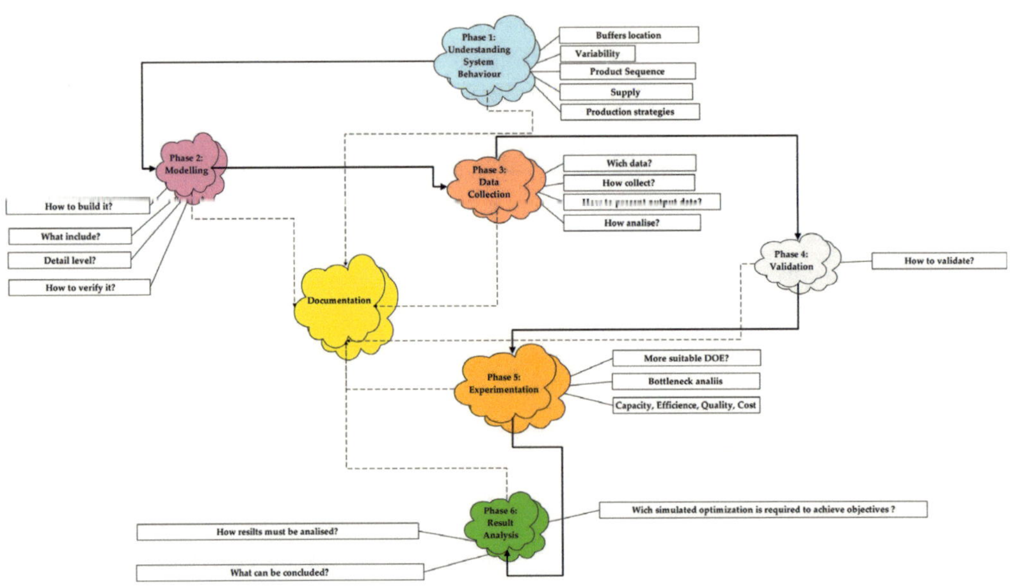

Figure 3. Methodology proposed by Banks and Tako for simulation adapted to SFS. ([45] p. 3528).

Phase 1: Understanding the problem. Understanding involved variables in system behavior is essential to propose an adjusted model regarding real system. It is important to understand variable appearing and disappearing, impact on production process and variability, even when production strategies are changed. This let to define clear objectives. Analyst must consider possible interest variables like: production time, assembly time, product quantity per time unit (capacity and efficiency), inputs and raw materials consumed per time unit, buffering time, etc. in this phase, a LF is a "Test Bank" to analyze how those interesting variables impacts plant layout.

Phase 2: Digital model building. Digital model main purpose is to represent adequately variables behavior. Sometimes it is necessary to make assumptions to simplify or complicate it, so there must have an adequate balance between easy use and high accuracy. Model must be detailed enough to capture as much information as possible according to study objectives. It is important to define detail level and construction method. LFs have limited size, and they allow build not very robust digital models with a high detail level and its construction does not demand excessive time. This aspect is an attractive alternative for future research because currently have low publications.

Phase 3: Data Collection: Information quality, and not quantity, allows more accurate analysis. LFs flexibility let to analyst adapt plant layout to data acquisition needs, considering product assembly sequence. In other words, relationship between production model and its critical variables differ from one type of product to another, however, the LFs can be adapted to certain types of products for data collection.

Phase 4: Model Validation. Validate a model is to obtain high degree of confidence to ensure correct predictions about process being simulated. There is no way to know which is most correct process to validate a model, however, experts consider relevant some guidelines for develop of a valid and credible model [31]:

- Use quantitative techniques to validate model components.
- Carry out sensitivity analyzes to determine most important factors in the model.
- Review simulation results to verify if they seem coherent. Here, again, LFs can be used. Given their resources redistribution availability, they are adequate to validate different simulated plant distribution models, becoming an interesting place for Engineers, Technicians and Managers, to evaluate decisions making impacts on productive practices.

Phase 5: Experimentation: Main experiments objective is to check whether the model complies with the assumptions and simplifications considered for validation. A good Design of Experiments (DOE) is crucial at this stage. It consists in determining tests and methodologies should be performed to obtain data and provide objective statistically evidence to answer the questions raised in phase 1 and clarify those process uncertain aspects that are being simulation object [32]. Once again, the LFs resources redistribution availability can be used to execute several experimental designs, expanding data collection opportunities portfolio.

Phase 6: Results Analysis. Finally, simulation results must lead to an appropriate decision making, either analyst consider that model is reliable and can use it to estimate the real one productive process variations or to discard it. Last case implies a new approach, a new analysis, and therefore anew simulation and a new digital model.

5. Shop floor simulation in industrial and academics scenarios

Layout planning can be categorized by static or dynamic environments. In the static layout approaches, material flow between machines is constant and an optimum layout is designed for a single time. On the contrary, if layout is evaluated and modified occasionally with respect to changes, research is categorized as dynamic. There are several plant simulation and discrete event simulation tools researching published, some of which have proposed design methodologies to structure, control and improve manufacturing processes [33, 34]. Below are some of them, both in industrial plants and training environments. The review of these cases cannot be considered complete but serves as a basis to demonstrate the capabilities of SFS in different scenarios.

5.1. Shop floor simulations in industrial production plants

Simulation based plant design and optimization techniques at industrial level have been implemented and developed several years ago. However, as mentioned in Section 1, there are many circumstances where uncertainty, interactions, objective functions or system constraints cannot be described using mathematical terms. In these cases, both, heuristic approach and SFS are often used to minimize some goals variables. For large dynamic plant layout problems, for example, some authors have suggested genetic algorithms-based heuristics solution methods [35].

Recently Pourhassan et al. [36] have presented mathematical models for relevant costs optimization during arrangement and re arrangement manufacturing facilities using genetic algorithms. A manufacturing system consisting of m machines processing n different products was considered. Products require m machines processing subsets causing material flow between they and eventually causing interference during material travel. Transporters are responsible to move materials between machines and there is plan divided into a several periods. Rate demand for each period is predicted and so material flow matrix is known in advance. Due to layout performance regarding to workflow is determined by number of possible transporters interference, researchers focused in to minimize material handling cost (MHC) and workflow interference as two inconsistent objective function:

$$Min\ Z_1 = \sum_{t=2}^{T}\sum_{i=1}^{N}\sum_{j=1}^{N}\sum_{l=1}^{N} A_{tijl}*Y_{tijl} + \sum_{t=1}^{T}\sum_{i=1}^{N}\sum_{j=1}^{N}\sum_{k=1}^{N}\sum_{l=1}^{N} F_{tik}*D_{jl}*C_{ik}*X_{tij}*X_{tkl} \quad (1)$$

$$Min\ Z_2 = Number\ of\ possible\ transporters\ accident \quad (2)$$

This model is subjected to decision variables:

$$X_{tij} = \begin{cases} 1, \text{if machine } i \text{ is allocated to location } j \text{ in period } t \\ 0, \text{ otherwise} \end{cases}$$

and:

N = Number of machines/locations.

T = Number of periods in planning.

i,j,k,l = Index of machines/locations.

t = Index for time periods.

A_{tijl} = Cost to shift machine i from location j to l in period t.

F_{tik} = Total flow between machine i and k in period t.

D_{jl} = Distance between machine i to l.

$$\sum_{i=1}^{N} X_{tij} = 1, \forall i = 1, 2, \ldots N, \forall t = 1, 2, \ldots T \tag{3}$$

$$\sum_{i=1}^{N} X_{tij} = 1, \forall j = 1, 2, \ldots N, \forall t = 1, 2, \ldots T \tag{4}$$

$$Y_{tijl} = X_{(t-1)ij} * X_{til} \forall i, j, l = 1, 2, \ldots .N, \forall t = 2, \ldots T \tag{5}$$

Constraint Eq. (4) ensures that each machine should be in one position and constraint Eq. (5) ensures that in each position only one machine should be allocated.

Function Eq. (2) was evaluated through simulation models where a possible finite set of layout scenarios were reached using Design of Experiments (DOE). In this case, a case study with five machines in a vehicle part production system with a planning horizon of two periods was considered. This allowed researchers to find by regression a Z2 optimal function (a minimal number of possible transporters accident). This work shows how numerical simulation, DOE and regression can be applied to evaluate a manufacturing system and recommends considering dynamic layout problem where the flow between machines is stochastic.

So, it is important to study if LFs can reach this situation, like the one described above, in which material flow between machines is constant (static design) or modified occasionally with respect to changes (dynamic design). Again, flexibility of LFs can be useful for this purpose.

On the other hand, there are several investigations using only SFS for plan layout planning to arrange and re arrange plant distribution. Filip et al. [37], for example, used Witness® software to optimize production systems in a printed circuit board (PCB) assembly plant. After completing initial simulation process and a series of subsequent simulations, researchers were able to identify equipment with highest workload, and part quantity produced. They also determined most economical method to increase production, which was to include new intermediate buffers. After including the new buffers, overloaded machine workload was reduced by 16.7%, resulting an increase of 1804 finished parts. This is an example about how understanding problem and setting clear objectives can lead to very successful results using SFS, an interesting aspect to be considering in LFs implementation.

Siderska [38] used Tecnomatix Plant Simulation® to simulate plant distribution and logistics in wire steel production process. Because of simulation, process statistics showed that only 10% of cutting machines capacity was being used due to a bottleneck generated by tip forming machine. An improvement was suggested simulating an additional tip forming machine in parallel with the initial one, increasing production. However, no subsequent work to was published by researchers to confirming if proposed change was implemented in the real plant, so simulated model could not be validated.

Hnát et al. [39] implemented Tecnomatix Factory CAD/Factory Flow in Zilina Intelligent Manufacturing System (ZIMS) laboratory, at Central European Technologies Institute (CEIT) for logistics process design and control. They described how these tools can be used for logistics systems conceptual design, control and monitoring. These researchers emphasize that using these technologies, companies can avoid failures due to inadequate design of material flows using simulation to verify flow before implementation.

Kikolski [40] applied simulation models to study different production scenarios. With Tecnomatix Plant Simulation, author simulated a real production plant using known process technological data and related material flow data to production. Their experiments analyzed how batch size influences system efficiency. The first simulated scenario consisted of the current conditions of the plant, whose lots comprise series of 30 items and each component in the system. For Kikolski, simulation model is an excellent tool to verify processes performance and give a clear visualization about assumptions. Analyzes require availability of one initial already system to design and optimize their virtual models. Therefore, it is observed that has been proposed start from initial distribution by other authors in other environments than the LFs one. Kikolski insists about continuing work on modeling and simulation methods, since "computer simulation can become a useful and reliable tool to design and study manufacturing processes and can provide a basis for other studies aimed at using digital models within of production engineering".

Kliment et al. [41, 42] have worked with Tecnomatix Plant Simulation to simulate logistics processes and production process on several production plants. Again, these works have focused on problems detection before physical plant building, and no evidence was found indicating proposed changes implementation during the simulation stage, so actual plant validation does could not be confirmed.

In Colombia, production plants simulation studies have been carried out in specific sectors such as mining [43] and biofuels [44], but do not in manufacturing plants, do not reach an appreciable reproducibility and scalability level and not involve entire value chain since raw materials supply to distribution chain. In conclusion, although there is an international trend, in Colombia there are no published works in which production plants simulation has been validated along value chain, or their knowledge has been transferred to industrial sector by academic institutions and much less by LFs.

5.2. Shop floor simulations in educational environments

About educational environments, an experience to highlight is in Skövde University [45]. This institution has been able to demonstrate impact and importance of plants layout simulation on decision making. Its practical courses have competences focused on undergraduate students, engineers, technicians and managers, and contemplates strategies based on six simulation phases presented in Section 4.

Due to this training, there is more people using SFS to test new product variants and production flows in their companies and these people are agree about this training methodologies should be facilitated by software tools that can support fast modeling and advanced experiments in a practical way. This experience is an interesting alternative to be replicated in LFs and is an accolade to SFS use for plant layout design.

6. Conclusions and further research

Several authors have illustrated how SFS based production planning and control can be raised to business-level activities [46], therefore, decision making skills acquisition at all company levels becomes more important. Thus, EAFIT University Production Technologies Research Group (GITP) has started a LF implementation to become a focused digital LF towards Industry 4.0 development. In this LF, SFS will supports decision making in production planning and control; It intends become an innovation transmitter agent to students, technicians, engineers and managers. In this way, GITP proposes to establish a current state in SFS applied in LF to set medium term clear objectives, and as have been seen, information studied allows concluding:

1. SFS is an extensive research subject, and with enough technological maturity level to evaluate production systems behavior with digital validations. However, lack published works using SFS in LFs comprises a knowledge gap needed to be addressed.

2. There are not validated LFs digital models linking cause/effect relationship between production process and its critical variables.

3. Advantages mentioned in present work let to LFs be a suitable scenario for research in SFS towards Industry 4.0 digital plant. Given their size, flexibility and resources redistribution availability, they allow digital models construction and experiments running, and can be used for SFS implementation as a decision-making tool for production control and planning.

GITP has initiated experiential learning methodologies implementation in EAFIT University LF [47], and according to Guarín and Baena [48] "experiential learning is most used pedagogical development mechanism (…). However, not enough research has been done to systematically address experiential pedagogical structure in LF". Thus, and due to some authors have classified simulation as an experiential learning method [49], there is also a high interest in evaluating SFS role in teaching/learning process. This subject will be addressed in future research once LF digital model have been constructed.

GITP considers EAFIT university LF digital model construction is needed. This digital model must be supported with good DOE to validate production system behavior, either by stochastic methods as described in Section 5.1 or SFS software. Initially, it is proposed to define a product and implement its assembly in LF production line. A data collection at process initial state will let define critical variables which have more impact in time, cost and production quality. Later, using DOE, alternatives for plant redistribution will be proposed to optimize

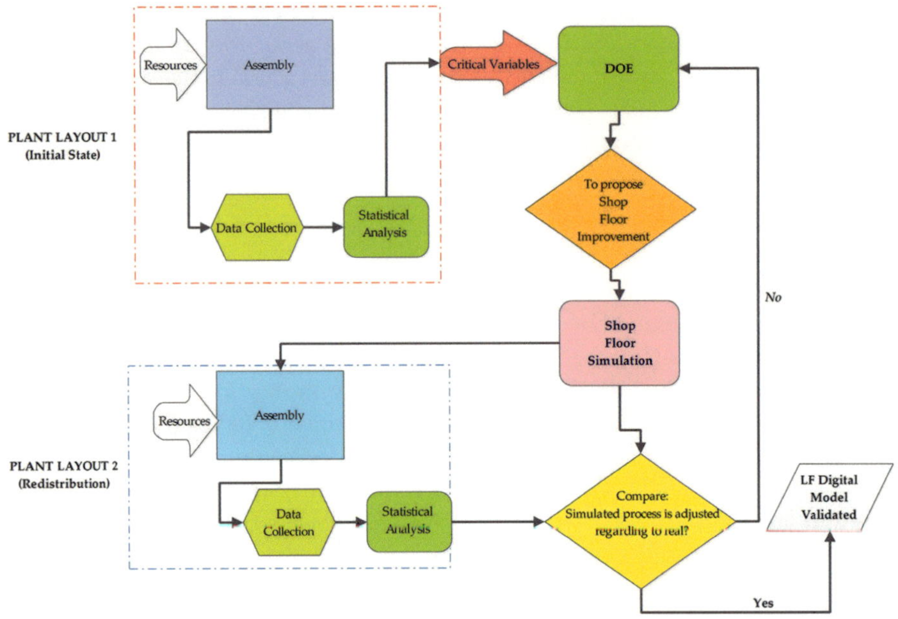

Figure 4. Proposed methodology for based SFS digital model construction in EAFIT University LF.

critical variables and verify statistically system response. These alternatives and information collected will be fed to an SFS software. Results obtained will be implemented by redistributing LF to validate digital model. See **Figure 4**.

As can be seen, there are several ways to take Learning Factories advantage for production activities research towards Industry 4.0, and shop floor simulation use and teaching for decision-making is one of the most relevant aspects to be investigated in the medium term, this is main reason to prepare this chapter.

Author details

Montoya Andrés*, Guarín Álvaro and Mora Julián

*Address all correspondence to: afmontoyav@eafit.edu.co

EAFIT University, Medellín, Colombia

References

[1] Hari Prasad N, Rajyalakshmi G, Sreenivasulu Reddy A. A typical manufacturing plant layout design using CRAFT algorithm. Procedia Engineering. 2014;**97**:1808-1814. DOI: 10.1016/j.proeng.2014.12.334

[2] Veres P, Bányai T, Illés B. Optimization of in-plant production supply with black hole algorithm. Solid State Phenomena. 2017;**261**:503-508. DOI: 10.4028/www.scientific.net/SSP.261.503

[3] Dengiz B, Bektas T, Ultanir AE. Simulation optimization based DSS application: A diamond tool production line in industry. Simulation Modelling Practice and Theory. 2006; **14**(3):296-312. DOI: 10.1016/j.simpat.2005.07.001

[4] Shannon R, Johannes JD. Systems simulation: The art and science. IEEE Transactions on Systems, Man, and Cybernetics. 1976;**SMC-6**(10):723-724. DOI: 10.1109/TSMC.1976.4309432

[5] Gaete-Quezada RA. University education: role playing as a strategy for evaluating university learning. Educación y Educadores. 2011;**14**(2):289-307. DOI: 10.5294/edu.2011.14.2.3

[6] Botero TS. Aplicación de la Realidad Virtual en la enseñanza de la ingeniería de la construcción. [thesis]. Universidad EAFIT; 2013

[7] Lucas R. Las Simulaciones de Empresa: una potente herramienta de aprendizaje. rrhh-Magazine.com [Online]. 2004. Available: http://www.rrhhmagazine.com/articulos.asp?id=282 [Accessed: 28-March-2018]

[8] Gutiérrez Fernández M, Romero Cuadrado MS, Solórzano García M. El aprendizaje experiencial como metodología docente: aplicación del método Macbeth. Argos. 2011; **28**(54):127-158

[9] Cachay J, Wennemer J, Abele E, Tenberg R. Study on action-oriented learning with a learning factory approach. Procedia - Social and Behavioral Sciences. 2012;**55**:1144-1153. DOI: 10.1016/j.sbspro.2012.09.608

[10] Abele E et al. Learning factories for research, education, and training. Procedia CIRP. 2015;**32**(Clf):1-6. DOI: 10.1016/j.procir.2015.02.187

[11] Tisch M, Metternich J. Potentials and limits of learning factories in research, innovation transfer, education, and training. Procedia Manufacturing. 2017;**9**:89-96. DOI: 10.1016/j.promfg.2017.04.027

[12] WU S-YD, WYSK RA. An application of discrete-event simulation to on-line control and scheduling in flexible manufacturing. International Journal of Production Research. 1989; **27**(9):1603-1623. DOI: 10.1080/00207548908942642

[13] Rogers PJGR. Simulation for real-time decision making in manufacturing systems. In: Proceedings of 1993 Winter Simulation Conference; 1993. pp. 866-874. DOI: 10.1145/256563.256863

[14] Lucke D, Constantinescu C, Westkämper E. Smart factory—A step towards the next generation of manufacturing BT. In: Manufacturing Systems and Technologies for the New Frontier. London: Springer London; 2008. ISBN: 978-1-84800-267-8

[15] Smith JS, Wysk RA, Sturrock DT, Ramaswamy SE, Smith GD, Joshi SB. Discrete event simulation for shop floor control. In: Proceedings of Winter Simulation Conference No. August 2015; 1994. pp. 962-969

[16] Son YJ, Joshl SB, Wysk RA, Smith SJ. Simulation based shop floor control. Journal of Manufacturing Systems. 2005;**2**(5):380-394

[17] Rao Y, He F, Shao X, Zhang C. On-line simulation for shop floor control in manufacturing execution system. Lect. Notes Comput. Sci. (including Subser. Lect. Notes Artif. Intell. Lect. Notes Bioinformatics), Vol. 5315 LNAI. No. PART 2.2008. pp. 141-150. DOI: 10.1007/978-3-540-88518-4_16

[18] Dias LMS, Vieira AAC, Pereira GAB, Oliveira JA. Discrete simulation software ranking—A top list of the worldwide most popular and used tools. In: Proceedings—Winter Simulation Conference; Vol. 2015. 2017. pp. 1060-1071

[19] Aceituno PA. La flexibilidad de la producción: sistemas flexibles de fabricación y de gestión de inventarios. BLOGS udima [Online]. 2018. Available: https://blogs.udima.es/administracion-y-direccion-de-empresas/libros/introduccion-a-la-organizacion-de-empresas-2/unidad-didactica-5-el-sistema-de-produccion-de-la-empresa/4-la-flexibilidad-de-la-produccion-sistemas-flexibles-de-fabricacion-y-de-gestion-d [Accessed: 01-May-2018]

[20] Dassault Systems. 3D Experience Lab [Online]. Available: https://3dexperiencelab.3ds.com/en/projects/fablab/3dexperiencer-center [Accessed: 28-March-2018]

[21] Henning M, Hagedorn-Hansen D, Von Leipzig K. Metacognitive learning: Skills development through gamification at the stellenbosch learning factory as a case study. The South African Journal of Industrial Engineering. 2017;**28**(3):105-112. DOI: 10.7166/28-3-1845

[22] Chryssolouris G, Mavrikios D, Papakostas N, Mourtzis D, Michalos G, Georgoulias K. Digital manufacturing: History, perspectives, and outlook. Proceedings of the Institution of Mechanical Engineers, Part B: Journal of Engineering Manufacture. 2009;**223**(5):451-462. DOI: 10.1243/09544054JEM1241

[23] Abele E, Cachay J, Heb A, Scheibner S. (2011) 1st Conference on Learning Factories - May, 19th 2011, Darmstadt. Institute of Production Management, Technology and Machine Tools (PTW), Darmstadt

[24] Abele E et al. Learning factories for future oriented research and education in manufacturing. CIRP Annals. 2017;**66**(2):803-826. DOI: 10.1016/j.cirp.2017.05.005

[25] Celar S, Turic M, Dragicevic S, Veza I. Digital Learning Factory at FESB – University of Split. 2016. [Online]. Available: https://bib.irb.hr/datoteka/806849.Celar_YUINFOICIST_2016.pdf

[26] Kesavadas T. V-Learn-Fact: A new approach for teaching manufacturing and design to mechanical engineering students. In: ASME 2013 International Mechanical Engineering Congress and Exposition; 2013. pp. V005T05A037-V005T05A037

[27] Matt DT, Rauch E, Dallasega P. Mini-factory—A learning factory concept for students and small and medium sized enterprises. Procedia CIRP. 2014;**17**:178-183. DOI: 10.1016/j.procir.2014.01.057

[28] De Jesus A, Grisales G, Restrepo FB, Orozco JM. Fábrica de Aprendizaje: Nuevo modelo de enseñanza productiva. In: Proceedings of the International Conference on Industrial Engineering and Operations Management; 2017. pp. 829-838

[29] Jerry B. Discrete Event System Simulation. Pearson Education India; 2005. ISBN: 813175 8966

[30] Tako AA. Model development in discrete-event simulation: Insights from six expert modelers. In: Proceedings—Winter Simulation Conference; No. December 2011; 2011. pp. 3923-3934. DOI: 10.1109/WSC.2011.6148083

[31] Blesa AG, Camio CA, Giménez FM. Desarrollo y Validación de un Modelo de Simulación para el Complejo Asistencial Médico Tecnológico de Navarra (CAMTNA); 2010

[32] Pulido HG, de la Salazar RV. Análisis y Diseño de Experimentos. 2nd ed. México DF: McGraw Hill; 2008. ISBN: 9789701065266

[33] Pavlović M, Arsovski S, Arsovski Z, MIROVIć Z, LAZIć M. Design methodology for discrete event simulation solutions in manufacturing environment. Strojarstvo. 2011;**53**(2): 113-126

[34] Mirović Z. Quality through integration of production and shop floor management by discrete event simulation. International Journal for Quality Research. 2007;**1**(2):132-138

[35] Balakrishnan J, Cheng CH, Conway DG, Lau CM. A hybrid genetic algorithm for the dynamic plant layout problem. International Journal of Production Economics. 2003; **86**(2):107-120. DOI: 10.1016/S0925-5273(03)00027-6

[36] Pourhassan MR, Raissi S. An integrated simulation-based optimization technique for multi-objective dynamic facility layout problem. Journal of Industrial Information Integration. 2017;49-58. DOI: 10.1016/j.jii.2017.06.001. ISSN: 2452414X

[37] Filip F-C, Marascu-Klein V, Mebrahtu H, Deaky B. Method for optimization production systems by computer aided modeling and simulation. Journal of Electrical and Electronic Engineering. 2012;**5**(1):67

[38] Siderska J. Application of Tecnomatix Plant Simulation for Modeling Production and Logistics Processes. Business, Management Education. pp. 64-73, 2016. DOI: 10.3846/bme.2016.316. ISSN: 2029-7491

[39] Hnát J, Herčko J, Gregor M. Design and Control of Logistic Processes based on Industry 4.0 Principles [Online]. 2016. Available: https://www.researchgate.net/publication/29935 4341 [Accessed: 05-May-2018]

[40] Kikolski M. Study of production scenarios with the use of simulation models. Procedia Engineering. 2017;**182**:321-328. DOI: 10.1016/j.proeng.2017.03.102

[41] Kliment M, Popovič R, Janek J. Analysis of the production process in the selected company and proposal a possible model optimization through PLM software module tecnomatix plant simulation. Procedia Engineering. 2014;**96**:221-226. DOI: 10.1016/j.proeng.2014.12.147

[42] Kliment M, Trebuňa P. Simulation as an appropiate way of verifying the efficiency of production variants in the design of production and non-production systems. Acta Logistica—International Scientific Journal. 2014;**4**(1):17-21

[43] Bustamante Rúa MO, Daza Aragón AJ, Bustamante Baena P. Barros Daza MJ. Simulación de plantas de procesamiento de minerales a través de MODSIM®. Revista Boletín Ciencias de la Tierra, pp. 33-37, 2016. DOI: 10.15446/rbct.n39.50451. ISSN: 2357-3740

[44] Rafael De la Rosa Ramos L, Henríquez Montero E, Sánchez Tuirán E, Angélica Ojeda Delgado K. Diseño y simulación de una planta para la producción de biodiésel a partir de Jatropha curcas L. en el departamento de Bolívar [Online]. 2014. Available: http://www.scielo.org.co/pdf/rion/v28n1/v28n1a08.pdf [Accessed: 08-March-2018]

[45] Ng AHC, Frantzen M. Production simulation education using rapid modeling and optimization: Successful studies. In: Proceedings of the 2015 Winter Simulation Conference; Vol. 2011. 2015. pp. 3526-3537

[46] Lee S, Son Y-J, Wysk RA. Simulation-based planning and control: From shop floor to top floor. Journal of Manufacturing Systems. 2007;**26**(2):85-98. DOI: 10.1016/j.jmsy.2007.07.001

[47] Baena F. Integración de la Teoría de Aprendizaje Experiencial a la Fábrica de Aprendizaje [thesis]. Universidad EAFIT; 2017

[48] Guarín Á, Restrepo FB. Fábrica de aprendizaje, una propuesta Didáctica. In: Tecnología, innovación e investigación en los procesos de enseñanza-aprendizaje. Barcelona, 2016. ISBN: 978-84-9921-848-9

[49] Crawley E, Söre Östlund J, Brodeur D, Edström K. Rethinking Engineering Education. 2nd ed. Springer International; 2014. ISBN: 9781612843926

Chapter 3

Impact of Industry 4.0 on Inventory Systems and Optimization

Xue-Ming Yuan

Additional information is available at the end of the chapter

http://dx.doi.org/10.5772/intechopen.90077

Abstract

With evolution of Industry 4.0, how should we operate our production lines and factories, how should we manage and optimize inventory, how should we deploy our workers, how should we run our businesses, how should we manage our supply chains? This chapter aims to highlight the impact of Industry 4.0 on manufacturing systems and services, as well as supply chains, in particular, on inventory systems and optimization. An integrative R&D framework for inventory systems modeling and optimization is proposed, which directs our R&D effort in modeling and optimizing inventory systems with Industry 4.0.

Keywords: Industry 4.0, cyber-physical systems, Internet of Things, Internet of Services, inventory systems, inventory optimization

1. Introduction

The world is experiencing Industry 4.0, the fourth industrial revolution. The first industrial revolution took place in the eighteenth century with the introduction of mechanical production machines powered by water and steam. The second industrial revolution started at the beginning of the twentieth century with mass production powered by electric energy. The third industrial revolution came in 1970s with production automation using electronics, computers and information technology. The current industrial revolution began in the early of this millennium with autonomous production using Cyber-Physical Systems (CPS), Internet of Things (IoT) and Internet of Services (IoS). This digitization not only enables the integration of processes and systems across companies and industrial sectors, but also creates new business models and value generation opportunities.

Industry 4.0 has been greatly influencing people's daily life in every aspect, from shopping to dining, from working to entertaining, etc. It is changing people's life styles and living behaviors, even thinking and mindset. Industry 4.0 has brought a revolutionary impact to manufacturing systems and services, as well as supply chains. In the environment of Industry 4.0, factories are smart, products are smart, and customers are demanding for being all round served with great satisfaction. Enterprises and businesses are digitalized, profitable and sustainable. Manufacturing systems and services are real time capable, interoperable, modular, decentralized, virtualized, and service oriented. Supply Chains are fully visible, connected and integrated.

Digitalization, visibility, connectivity and interoperability are the essence of Industry 4.0. With rapid growth of Industry 4.0 technologies, inventory systems and optimization are being transformed to a new state. This chapter explores what is the impact of Industry 4.0 on inventory systems and optimization, how Industry 4.0 enables the transformation of inventory systems and optimization to a new state and what are the benefits of such transformation to the industry.

We briefly review Industry 4.0 and enabling technologies in Section 2. In Section 3, we discuss the possible changes with factories, products, customers and businesses in the environment of Industry 4.0. Section 4 highlights the attributes of manufacturing systems and services with Industry 4.0. Section 5 discusses the benefits brought by Industry 4.0 to businesses. In Sections 6–8, we focus on inventory systems and optimization. Section 6 explores the impact of Industry 4.0 on inventory systems, and Section 7 discusses the impact of Industry 4.0 on inventory optimization. In Section 8, we propose a new integrative R&D framework for inventory systems modeling and optimization. Section 9 concludes the chapter.

2. Industry 4.0 and enabling technologies

Industry 4.0 is a transition to the digital transformation of industries, a merger of the physical and digital worlds. Industry 4.0 is also a fusion of technologies that clear the boundaries among the physical, digital, and biological spheres [1–3]. Those technologies include Artificial Intelligence, Robotics, Internet of Things, Autonomous Vehicles, 3-D Printing, Nanotechnology, Biotechnology, Materials Science, Energy Storage, and Quantum Computing [4–10]. In this chapter, we shall not detail the mentioned technologies, but briefly highlight how Industry 4.0 is enabled by the technologies.

Industry 4.0 is enabled by the technologies that integrate the digital and real worlds. As an illustration, the core technologies related to manufacturing systems and services, and supply chains, are elaborated as follows:

Big Data Analytics and Business Intelligence: Nowadays enterprises are embraced by huge amount of data from all kinds of sources. All decisions, either operational or managerial, have to rely on data analytics. All businesses cannot afford to ignore the data and data analytics. It has been proven that the data are very useful and much valuable when it comes to optimize production quality and services, reduce energy consumption, and improve efficiency in manufacturing. For example, data can be collected from various stages of the manufacturing process. The large amount of data can be analyzed in correlation with each other in order to identify the stages with redundant processes that need be streamlined. The data can be turned into actionable insights through early warning algorithms, predictive models, decision support, workflows and dashboards. In fact, the data have become the most valuable asset of an organization today. The data analytics is the essence to a success of enterprises and their businesses.

Internet of Things (IoT): Embedded computing and networking connect sensors and devices. The Internet of Things connects more and more systems, devices, sensors, assets and people through networks ranging from wireless, low-power wide-area networks to wired high-capacity networks.

Cloud Computing: Many small enterprises cannot afford to own the storage for an increasing amount of data. Some of them lack the analytical and computational capabilities of processing the huge amount of data. However, the data and the capabilities of analytics and computation are even more important to them. Cloud service providers are offering a growing opportunity to small enterprises through data storage and data processing with affordable costs.

Cyber-Security: Industrial systems are becoming increasingly vulnerable to threats as there are various devices and large amount of data sets. The cyber-security has to be put in place, which can recognize new vulnerabilities and challenges, and ensure industrial systems and businesses free of attacks.

Autonomous Robots: The use of robots in manufacturing is no longer new. In Industry 4.0, robots are self-sufficient, autonomous, and interactive, which are no longer simply tools used by humans, but are the integral work units.

Additive Manufacturing: Additive manufacturing, in particular, 3D printing enables manufacturers to dramatically reduce the design time and cost, and increase the variety and customization of products. The benefits and values brought by additive manufacturing are even more evident in the high mix low volume (HMLV) manufacturing.

Augmented Reality: Customers are now more and more demanding. Businesses have to provide customers all round services, from personalized product design, dedicated manufacturing process to individualized after-sales services. Augmented reality based systems are able to assist enterprises to gain their competitive edges and win more business opportunities in such a competitive marketplace.

3. Factories, products, customers and businesses in the environment of Industry 4.0

Industry 4.0 uses digital technologies to make manufacturing more agile, flexible and responsive to customers. It is able to create a smart factory where the Internet, wireless sensors, software and other advanced technologies work together to optimize the manufacturing system and improve customer satisfaction. Industry 4.0 enables a business to react more rapidly to market changes, offer more personalized products and increase operational efficiency in a cycle of continuous improvement.

Industry 4.0 is creating intelligent products, processes and procedures. In a smart factory, workers, machines and resources easily communicate via the ubiquitous connectivity of people, things and machines. Products, transportation equipment and tools cooperate in order to create better each following production step. It leads to the connectivity of virtual world and physical objects in the real world.

Factories: With Industry 4.0, not only all manufacturing facilities in a factory, such as sensors, actuators, machines, robots, conveyors, etc., are connected and exchange information automatically, but also the factory becomes conscious and intelligent enough to predict and maintain its operational performance, to control its production process, and to manage its manufacturing operations. In addition, manufacturing processes, such as product design, production planning, product services, etc., are controlled by a decentralized system but working fully independently as self-functional modular. Such a factory with Industry 4.0 is known as a smart factory.

Products: With Industry 4.0, a product is embedded with sensors, identifiable components, and processors which can carry information and knowledge to convey the functional guidance to the customers, and transmit the users' feedback to the manufacturing system. More importantly, it allows the manufacturer to monitor the product performance and provide information to customers, such as when and which parts need to be replaced, and generate more revenue by shifting to a product-as-a-service business model. Such a product with Industry 4.0 is named as a smart product. A smart product has all information about itself, manufacturing time, production conditions, delivery time, utilization, life time, location, and other important information [11].

Customers: Industry 4.0 provides customers a new purchase means with a lot of advantages. It allows customers to order whatever function of products, with any number even if only one. In addition, customers could change their orders and ideas at any time during production even at the last minute with no extra charge. On the other hand, the benefit from the smart products enables the customer not only to know the production information of the product but also to receive the advice of utilization depending on their own behaviors.

Businesses: With Industry 4.0, there is a complete communication network existing among various entities such as suppliers, factories, customers, products, logistics, resources, etc. Every entity optimizes its configuration in real-time, based on its demand and state in the network so as to maximize the profit for all the entities with the limited sharing resources, and at the same time, to reduce the costs and pollution, raw materials, CO_2 emissions, etc.

4. Manufacturing systems and services with Industry 4.0

Industry 4.0 is connecting systems, machines, and work units in order to create intelligent networks along the value chain that can work separately and control each other autonomously but in a cohesive manner. In the Industry 4.0 environment, the key attributes of manufacturing systems and services are real time capability, interoperability, modularity, decentralization, virtualization and service orientation.

Real-Time Capability: Industry 4.0 enables everything real time, which requires that the manufacturing process, data collecting, monitoring and maintenance are in a real time manner.

Interoperability: Interoperability refers to the capability of all entities to connect, communicate, and operate together via the Internet of Things. This includes the humans, the smart factories, and the relevant technologies. Interoperability requires an entire environment with smooth interaction and flexible collaboration among all the entities. For example, assembly stations are not separate or isolate from the products created or the people who are working on them.

Modularity: Modularity enables smart factories to easily adapt the changing circumstances and requirements. Modularizing products and manufacturing systems ensures a minimum disruption to other products or manufacturing processes when there is a need to replace, expand or improve an individual product or production line.

Decentralization: Industry 4.0 supports decentralization, which enables the different systems within the smart factory to make decisions autonomously, in alignment with the ultimate overall organizational goal.

Virtualization: A virtual twin or digital twin can be created for a smart factory to greatly enhance the existing processes and products, and reduce the time to profit of new products. What are actually happening in the physical factory is exactly mapped to the virtual digital factory. Engineers and designers can work on the virtual digital factory, then customize, alter, and test changes or upgrades in complete isolation, without affecting the physical factory operations.

Oriented Services: The Internet of Things creates potential services that customers desire to consume, which include those services within or outside a smart factory, or before or after sales of a product. The Internet of Services is an important component of Industry 4.0.

5. Benefits to businesses

Industry 4.0 is making it easier for companies to collaborate and share data among customers, manufacturers, suppliers and other parties in supply chain. It improves productivity and competitiveness, enables the transition to a digital economy, and provides opportunities to achieve economic growth and sustainability.

In the environment of Industry 4.0, all the parties in the supply chain share the data from their production sites, vehicles, warehouses and databases in real time. Real-time POS (point of sales) and inventory data are available to understand the business situation. Customer urgent orders can be attended timely with customer satisfaction. Condition and location of products are trackable and controllable. Product quality is better controlled. Inventory is better managed. Equipment settings are self-adjusted based on materials used, products being made and other ambient conditions. Mass-produced products are customized according to the needs of an individual customer. Equipment can be monitored remotely and malfunctions can be predicted accurately. Whatever business is, a fluid digital continuum is able to connect customers, suppliers, partners, production equipment and products throughout the lifecycle of the product and services. The benefits brought by Industry 4.0 to businesses are specifically summarized as follows:

Increased Business Competitiveness: Industry 4.0 enhances global competitiveness through cooperation and a confederation of firms. It can be seen that products will no longer be built by a worker in future, but by a robot or programmer.

Increased Productivity and Revenue: With the increase in efficiency, lowering of operational costs leads to increased revenue and profits. This also drives forward improvements in productivity. Industry 4.0 is one of the key drivers for companies' revenue increase and nations' GDP growth.

Optimized Manufacturing Processes: Smart factories are all connected. The necessity is a network connecting smart factories, smart products, and other smart production systems. Cyber-physical production systems enable factories and manufacturing facilities to quickly and properly react to the changes in customer demand levels, stock levels, machine defects, and unforeseen delays. Marketing intelligence, smart logistics and intimate customer services are also crucial in the entire value chain. The integration facilitates the establishment and maintenance of networks that create and add value. It could also mean the integration of new business models across countries and even across continents, making for a global network.

Accelerated Technology Development: Industry 4.0 provides a platform for the basis of further innovation with developing technologies. Manufacturing systems and services can be further developed. For example, with mobile phone applications, more and more developers are using open APIs to mash up applications, and looking into technologies that will be an improvement on the current GPS, RFID, NFC, and even accelerometer sensors embedded in the standard smartphone [12, 13].

Better Customer Service: Industry 4.0 is able to monitor real time customer feedback so as to provide better service to customers.

In a summary, Industry 4.0 enables a digitally integrated and intelligent value chain offering almost limitless possibilities. Industry 4.0 solutions improve operations efficiency, productivity, product quality, inventory management, asset utilization, time to market, agility, workplace safety and environmental sustainability. In the following sections, to be more specific, we highlight the impact of Industry 4.0 on inventory systems and optimization, and propose a new integrative R&D framework for inventory systems modeling and optimization in the Industry 4.0 environment.

6. Impact of Industry 4.0 on inventory systems

In the business world, inventory is very important, which is directly linked to cash and cash flow. Inventory appears everywhere, in a visible form or non-visible form. In manufacturing, there are raw material inventory, work in process (WIP) inventory, and finished goods inventory, which are all in a visible form. In communication, for example, bandwidth, server and memory card capacity can be considered as inventory, in a non-visible form. Thus, to efficiently and effectively managing inventory, either in a visible or non-visible form, is the winning formula to businesses.

In the context of supply chain, suppliers have raw material inventory, manufacturers have raw material inventory, work in process (WIP) inventory and finished goods inventory, distributors have semi-product inventory and finished goods inventory, retailers have finished goods inventory. In each stage, inventory need be kept so as to improve the satisfaction level of its downstream stage, reduce certain costs and ensure efficient and effective operations of the supply chain. It is not favorable to hold inventory in each stage because of carrying cost, cash retention, product depreciation, etc. Inventory optimization is to keep the minimal inventory to maximally fulfill the downstream demand, that is, to keep the right balance between the supply from the upstream and the demand from downstream [14–17].

The impact of Industry 4.0 on inventory systems can be summarized as four aspects: inventory process, inventory classification, inventory system parameters, and inventory system review.

Inventory Process: Industry 4.0 enables the purchase process and fulfillment process digitalized and automated. For orders to suppliers, based on real time information and data, inventory systems are able to automatically trigger the orders with the right quantity at the right time. For the fulfillment to customers, inventory systems are able to foresee when and how much each customer needs which type of product through data analytics and business intelligence.

Inventory Classification: ABC inventory classification is popularly used in inventory management. ABC inventory classification is to start with the products ranked by dollar value in descending order, and plot the cumulative dollar value in inventory versus the cumulative products in inventory. For example, Class A is referred to the set of those 20% products cumulated 80% dollar value in inventory, and Class C is the set of 50% products cumulated 10% dollar value in inventory. As such, Class B is the set of the remaining 30% products cumulated 10% dollar value in inventory. In the environment of Industry 4.0, products are smart products which are carrying the information on their locations and dollar values. It enables the automation of ABC inventory classification of smart products. The inventory classification profile is dynamically and automatically updated in real time.

Inventory System Parameters: For an inventory system, its key system parameters include supply lead-time, purchasing price, carrying cost, ordering cost, customer demand and product selling price, which collectively determine the dynamics of the inventory system and its performance. It is difficult to estimate some of these system parameters, such as supply lead-time, customer demand, etc. When modeling the inventory system, it is conventional to take a rough estimation due to the lack of information for calculating their exact values. For example, the shipping time from the United Stated to Singapore is about 2–4 weeks. It means that the supply lead-time from a supplier in the United States to a manufacturer in Singapore may vary between 2 and 4 weeks. There is a huge difference to the production plan of the manufacturer when the supply lead-time is 2 or 4 weeks. It has been very challenging for the manufacturer to gauge the exact time to receive what it has ordered. Besides the supply lead-time uncertainty, inventory management is also facing other challenges from suppliers, such as minimum order quantity (MOQ), dynamic pricing, etc. In the environment of Industry 4.0, all the parties and entities are able to share the information in real time. In the example, the manufacturer in Singapore will be informed timely where the vessel is and when it will reach in Singapore. Then the manufacturer can plan its production accordingly at the right time. In this case of the supply lead-time parameter, the impact of Industry 4.0 on inventory system parameters has been evident. Industry 4.0 has similar effects on other inventory system parameters as well.

Inventory System Review: Conventionally there are two types of inventory reviews: periodic review and continuous review. The periodic review is to review the inventory on a regular basis, at fixed point of time, and a new order is placed only at these time epochs. The continuous review is to continuously monitor and review the inventory over the time, and a new order is placed when the inventory reaches a critical point. The periodic review does not need real-time information on the inventory while the continuous review needs real-time information about the inventory. That is, the continuous review needs a sophisticated advanced information system in place. Due to the high cost involved in the advanced information system, the most companies are adopting the periodic review in practice. However, with Industry 4.0, all the information is available and shared in a real-time manner. It provides the most convenience and great incentive for companies to start using the continuous review. Thus, Industry 4.0 is shifting the periodic inventory review to the continuous inventory review in practice.

7. Impact of Industry 4.0 on inventory optimization

The purpose to keep inventory is to buffer the uncertainties which may come from the upstream (e.g., suppliers) and the downstream (e.g., customers) so as to timely fulfill customer demand if any. However, to keep the inventory too high will incur a higher inventory cost, while keeping the inventory too little will compromise the customer satisfaction level. It is utmost important to keep the right inventory at the right time in the right place with the right price and the right time duration. Thus, inventory must be optimized to minimize the inventory cost and maximize the customer service level.

Inventory optimization is to decide when to order and how much to order, which constitute an inventory policy. The optimal inventory policy is determined by solving an optimization problem that is composed of an objective function and a set of constraints. The objective function and constraints define the relationships of the system parameters. The objective is either to minimize the total operational cost, or to maximize the customer service level. The decision variables are the time to place an order and the quantity of an order. The optimization problem is formulated based on the assumption of the system parameters. Some of the system parameters are constant, some are variables which change over the time, and some are random variables which change according to a certain probabilistic distribution. The main challenges to optimize an inventory system are how to accurately characterize the system parameters, how to formulate the relationships of the system parameters and construct the objective function and constraints, and how to derive the optimal solution to the optimization problem.

For an inventory system in the environment of Industry 4.0, the values of some system parameters are directly captured in the information systems [18–20]. Through data analytics, these system parameters can be well modeled and characterized. For some system parameters which are not directly recorded in the information systems, they can be analyzed and described based on their relations to other system parameters. It is very difficult to analytically formulate the relationships of the system parameters, in particular, when the inventory system is complex. With all available data and analysis, it might not be necessary to come out with mathematical formulas for the system parameter relationships. Through extensive deep data analytics, the optimal inventory policy is expected to be achieved as well.

8. New R&D framework for inventory systems modeling and optimization

Industry 4.0 enables digitalization, visibility, connectivity and interoperability across supply chain. The impact of Industry 4.0 on inventory systems and optimization is huge. Industry 4.0 is shaping a new R&D paradigm for inventory systems and optimization. As an initial attempt, we are proposing a new integrative R&D framework for inventory systems modeling and optimization in this section.

There are various types of inventories, in a visible or non-visible form. As an illustrative example, we consider the finished goods inventory in supply chain, where the upstream is suppliers and the downstream is customers, end-users of products, as shown in **Figure 1**.

The primary function of the inventory is to purchase the products from the suppliers and sell the products to the customers. Through such trading, buy and sell, the revenue will be generated. The main purpose to manage the inventory is to maximize the sales to the customers by holding the minimal stock on hand. Thus, inventory management need to clearly know its customers and well understand its suppliers. Industry 4.0 provides all opportunities for the inventory management to achieve its ultimate goal, maximizing its revenue.

To clearly know the customers is the first important to managing the inventory. Customer is a king, no customer no sales, no sales no revenue, no revenue no business. Through data analytics and market intelligence, customers' behavior is modeled, and future customer demand is forecasted. The product selling price might have great influence on customer demand. Similarly, the customer satisfaction affects customer demand as well. The analysis on the customer sensitivity to pricing and the effect analysis of the customer satisfaction are able to improve the accuracy of future customer demand forecasts.

To well understand the suppliers is able to reduce the operational cost in managing the inventory. Leveraging all available data and business intelligence, the supplier performance is evaluated and analyzed. From the performance analytics, it is ready to know that which

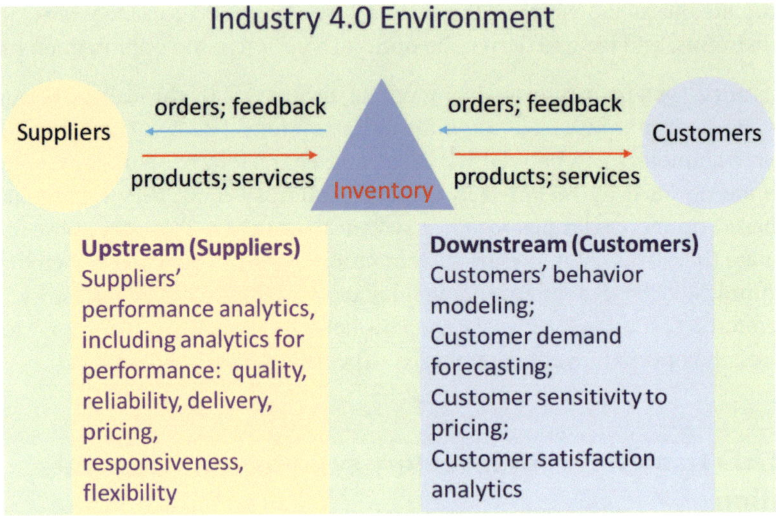

Figure 1. Integrative R&D framework for inventory systems modeling and optimization.

supplier can provide the best quality of products or the best service at which price, which supplier is most reliable, which supplier can deliver the products timely, and which supplier is most responsive and flexible to attend the last minute urgent order.

Conventionally, after estimating the system parameters and forecasting the customer demand, an inventory optimization model is built up to derive the optimal inventory decision in terms of the objective function. With Industry 4.0, all the data about the suppliers, customers and the inventory itself are available to be utilized for establishing an integrative data driven inventory optimization model. Instead of the conventional sequential approach with the assumptions on the system parameters, an integrative data driven approach is applied without the assumption on the system parameters. By knowing the customers and understanding the suppliers, the inventory can be managed efficiently and effectively so that the maximal revenue can be achieved with the maximal customer satisfaction.

9. Conclusions

Industry 4.0 enables factories smart, products smart, and supply chains smart as well, and makes manufacturing systems and services more agile, flexible and responsive to customers. Through a brief overview on Industry 4.0 and enabling technologies, this chapter discussed the possible changes with factories, products, customers and businesses in the environment of Industry 4.0. The attributes of manufacturing systems and services with Industry 4.0 were highlighted, and the benefits brought by Industry 4.0 to businesses were discussed. To be more specific, the chapter focused on inventory systems and optimization. The impact of Industry 4.0 on inventory systems and optimization was explored, respectively. The new integrative R&D framework for inventory systems and optimization was proposed in this chapter.

How to efficiently and effectively manage inventory is a common challenge for all businesses and companies. It has been a long standing issue in industrial practice, and there is no universal solution to all businesses and companies. It is probably because the conventional approaches and methods for inventory systems modeling and optimization have their limits, or there is lack of the information on knowing customers and understanding suppliers. With Industry 4.0 implementation and progress, it is anticipated that there will be more and more breakthroughs in approaches and methods for inventory systems modeling and optimization.

Author details

Xue-Ming Yuan

Address all correspondence to: xmyuan@simtech.a-star.edu.sg

Singapore Institute of Manufacturing Technology, Agency for Science, Technology and Research (A*STAR), Innovis, Singapore

References

[1] CGI Group Inc. Industry 4.0: Making your business more competitive. In: White Paper. CGI Group Inc. 2017

[2] PwC. Industry 4.0: Building the digital enterprise. In: White Paper. PwC. 2016

[3] Renjen P. Industry 4.0: Are you ready? Deloitte Review. 2018;**22**:8-11

[4] Abdel-Basset M, Manogaran G, Mohamed M. Internet of Things (IoT) and its impact on supply chain: A framework for building smart, secure and efficient systems. Future Generation Computer Systems. 2018;**86**:614-628

[5] Bartodziej CJ. The Concept Industry 4.0: An Empirical Analysis of Technologies and Applications in Production Logistics. Wiesbaden: Springer; 2017

[6] Bi Z, Xu LD, Wang C. Internet of Things for enterprise systems of modern manufacturing. IEEE Transactions on Industrial Informatics. 2014;**10**:1537-1546

[7] Boyes H, Hallaq B, Cunningham J, Watson T. The Industrial Internet of Things (IIoT): An analysis framework. Computers in Industry. 2018;**101**:1-12

[8] Crnjac M, Veža I, Banduka N. From concept to the introduction of Industry 4.0. International Journal of Industrial Engineering and Management. 2017;**8**:21-30

[9] Gilchrist A. Industry 4.0: The Industrial Internet of Things. New York City: Apress; 2016

[10] Ng IC, Wakenshaw SY. The Internet-of-Things: Review and research directions. International Journal of Research in Marketing. 2017;**34**:3-21

[11] Porter ME, Heppelmann JE. How smart, connected products are transforming competition: Spotlight on managing the Internet of Things. Harvard Business Review. 2014;**92**:64-88

[12] Leung J, Cheung W, Chu S-C. Aligning RFID applications with supply chain strategies. Information Management. 2014;**51**:260-269

[13] Wamba SF. Achieving supply chain integration using RFID technology: The case of emerging intelligent B-to-B e-commerce processes in a living laboratory. Business Process Management Journal. 2012;**18**:58-81

[14] Porteus EL. Stochastic inventory theory. In: Heyman DP, Sobel MJ, editors. Handbook in Operations Research and Management Science, Vol. 2: Stochastic Models. North-Holland; 1990. Ch. 12

[15] Prak D, Teunter R, Syntetos A. On the calculation of safety stocks when demand is forecasted. European Journal of Operational Research. 2017;**256**:454-461

[16] Silver EA, Pyke DF, Thomas DJ. Inventory and Production Management in Supply Chains. New York: Taylor and Francis; 2017

[17] Wu L, Yue X, Jin A, Yen DC. Smart supply chain management: A review and implications for future research. International Journal of Logistics Management. 2016;**27**:395-417

[18] Huber J, Muller S, Fleischmann M, Stuckenschmidt H. A data-driven newsvendor problem: From data to decision. European Journal of Operational Research. 2019;**278**:904-915

[19] Kleywegt AJ, Shapiro A, Homem-de Mello T. The sample average approximation method for stochastic discrete optimization. SIAM Journal on Optimization. 2002;**12**:479-502

[20] Levi R, Roundy RO, Shmoys DB. Provably near-optimal sampling-based policies for stochastic inventory control models. Mathematics of Operations Research. 2007;**32**:821-839

Cyber-Physical System Architecture for Minimizing the Possibility of Producing Bad Products in a Manufacturing System

Salvador Cobos Guzman, Maria Dolores Cima Cabal,
Francisco Machio Regidor and Lucia Alonso Virgos

Additional information is available at the end of the chapter

http://dx.doi.org/10.5772/intechopen.88461

Abstract

The new industry 4.0 requires the implementation of several cyber-physical systems to increase the level of productivity in a manufacturing system. This chapter proposes an architecture of a generic manufacturing system that requires the use of techniques of agile production, lean manufacturing, and statistical approaches. The combination of the previous techniques will be implemented in the architecture proposed for minimizing the possibility of producing bad products. Thus, the cyber-physical system architecture proposed will optimize the overall system thanks to the implementation of intelligent modules and control strategies. Moreover, 10 proposed actions will be described in detail. These actions can be implemented in cyber-physical systems that take into account five levels.

Keywords: cyber-physical systems, industry 4.0, lean manufacturing, agile production, increase of quality

1. Introduction

The new industry 4.0 requires high levels of digitalization in order to process all the information that is generated in virtual representations or cyber versions of the physical processes. This cybernetic level will produce massive information stored in servers that can be used for future digital analyses that intelligent algorithms will use as an input to optimize the process and monitor the system (e.g., detecting anomalies). Thus, decision-making can be produced automatically by the intelligent algorithms implemented or by means of the collaboration of human experts depending on the type of engineering process.

Moreover, the new smart industry must have the capability of reacting to new changes in the market or to different requirements from the customers. This is where the concept of agile manufacturing [1] is used in order to generate new adaptations to the manufacturing process, keeping high quality in the products in shorter times or producing just in time.

Otherwise, lean manufacturing is the set of resources that help to eliminate waste. Waste is understood as all operations that do not add value to the finished product, to the service offered or different processes. The elimination of waste seeks to REDUCE production costs, material, human resources, stock, inventory, overproduction, lead time/waiting time, transport, and movements in order to INCREASE efficiency production tasks, quality, and overall customer satisfaction. Lean manufacturing can be summarized in four blocks:

1. Minimize waste.

2. Increase quality.

3. Produce flexibly.

4. Offer a system of continuous improvement.

Nowadays, there is the possibility of integrating these lean manufacturing operations with cyber-physical systems (CPS) to optimize the process. This integration is used in the new industry 4.0 for providing high levels of optimization.

The advantage of industry 4.0 is to revolutionize the management of systems to improve their application in a smart manner. The CPS is used to intercommunicate machine-machine or human-machine taking advantage of the possibility of managing a broad set of useful data in decision-making. This enables the system to be intelligent and flexible in different contexts.

This chapter proposes a cyber-physical system architecture that can be used in the applications of industry 4.0 that automates and improves the effectiveness of concepts such as Just-in-Time methodology. Additionally, seven types of waste (*mudas* in Japanese) are analyzed and offer ten possible actions to help its implementation.

The structure of the chapter is as follows. Section 2 presents basic concepts about the type of waste and according to the most relevant definitions. Section 3 presents a literature review about the architectures of cyber-physical systems and the general contribution of the CPS proposed. Section 4 presents the proposed actions of the CPS architecture proposed. Finally, conclusions and future work are presented in Section 5.

2. Types of waste

In manufacturing, "waste" or *mudas* is defined as any activity that consumes resources without adding any value to the manufactured product [2]. There are seven types of *mudas*: (1) overproduction, (2) inventory, (3) overprocess, (4) reprocess, (5) wait, (6) transportation, and (7) movement.

These *mudas* belong to the Just-in-Time (JIT) methodology. Just-in-Time is a method that tries what its name indicates. It can be summarized as an effective system in terms of times that take into account the capacity of workers/machines, workload, and resources. Moreover, it is used to organize workers and tools, and it is responsible for reducing waste [2].

2.1. Overproduction and inventory

The first *muda* is overproduction. There are times when companies need to meet an established program, even when there are setbacks. "Overproduce" implies producing above the needs of the market. Many companies overproduce to ensure they do not run out of stock in the face of any setback. In this way, they always comply with the specifications of the program.

Producing requires spending on raw materials and energy and spending on time and expenditure on storage. Therefore, overproduction spends all this, but in a less justified way. According to the seven *mudas* of lean manufacturing, efficient means to produce at a rate coinciding with the demand [3]. When it occurs at a time higher than the demand, unjustified extra hours may be necessary. It is not a solution to produce at a lower rate than the demand because when this happens, it could produce possible waiting times.

The second *muda* is the inventory. An "inventory" is defined as the number of materials or information that is above the minimum necessary for production. The purpose of lean manufacturing is to save inventory of finished parts because these parts cause storage, cleaning, and maintenance expenses unnecessarily. Quality errors are quite common when inventory levels are high [4].

2.2. Overprocess and rework

Overprocessing and reprocessing are additional production activities or services. It is understood as additional what the client does not perceive and, therefore, does not confer value to the manufactured product [2].

Repeated activities and processes cause temporary expenditure. This expense can be avoided by reorganizing, grouping, or simplifying repeated activities and operations. For this, it is necessary to make a global analysis of the manufacturing plant. Lean manufacturing usually uses the value stream mapping system (VSM) to visualize the entire process. The objective is to detect those unnecessary activities or processes [5].

It is recommended to generate a database that includes all available machines and resources, as well as to collect information about the number of operators, productivity, cycle time, batch change, efficiency of machinery, loss of efficiency in operations, production plans, time that a piece is in the factory since it is raw material until it is terminated, previous and next process being analyzed, flow of information among machines, and data on maintenance.

With this information, a map is produced that offers a global analysis of the activities or processes. This map is called the initial VSM. Through this map, it can locate and solve problems effectively and propose a new map as a final VSM.

2.3. Waiting

Waiting is the time lost while waiting for the pieces from one department to another, especially during automatic production. This lost time affects mainly in the worker's activity. This effect can be detected during the VSM and can be removed by SMED or the method of total productive maintenance (TPM) [6].

TPM seeks to guarantee confidence in the processes by offering an activity with zero defects, zero accidents, or zero waste. In order to do this, prevention actions are analyzed and determined. This analysis should be done with the help of the operators through the free maintenance of the machine. This maintenance includes data about cleaning, dirt prevention, identification of anomalies, possible problems, etc. to solve any defect and lengthen the life of the machine.

The SMED method recommends changing the matrix in a single minute. Its objective is to reduce the preparation times of processes to reduce manufacturing and delivery times, which can lead to delays. This method observes and separates internal operations from external operations. Internal operations are those that can only be carried out with the machine stopped. These operations are unlocking, changing, and locking a mold or a tool. External operations are those that can be performed with the device turned on. These operations are, for example, to approach materials, consumables, or parts to a machine and other pre-adjustment actions, such as checking the states of the machines [6]. As it happens with the VSM, for the SMED a study is needed for analyzing the time invested in each activity or process and the type of operation.

2.4. Transport and movement

Although the tasks of transport and movement are considered almost fundamental in a manufacturing company, there are times when they are not essential. It is necessary to minimize the expense of transportation and mobility of parts when they do not add value to the process because it implies a temporary loss and unnecessary fatigue to the operators.

3. Literature review of cyber-physical systems

Actually, there are cyber-physical system (CPS) architectures [7, 8] that include five stages or levels such as (1) connection, (2) conversion, (3) cyberspace, (4) cognition, and (5) configuration. Another CPS architecture for manufacturing processes can be seen in [9] that contains also five levels such as (1) measurement, (2) acquisition, (3) signal processing, (4) decision support, and (5) loop control. In this architecture, the levels from two to five use data cloud services to process each level. The problem with this architecture is that real-time applications are not recommended to control devices directly from the cloud because the network can generate delays and consequently perturbations in real-time control.

Another example of CPS architecture is proposed by [10] for a CNC system. This CPS uses as well five layers such as (1) equipment, (2) sensing, (3) network, (4) cognitive, and (5) control. The three first layers represent the physical space, and the last two layers represent the cyberspace (cognitive and control).

According to these architectures, we propose also a cyber-physical system architecture based on five levels as it is shown in **Figure 1**.

The first level starts with the level of connection that is important among different machines. The communication can use machine-machine (MM) protocols (e.g., MQTT protocol) and requires connections plug and play [7] and plug and produce that can be connected automatically. This level can use also other ways to identify people like what is proposed in [11] for human collaboration.

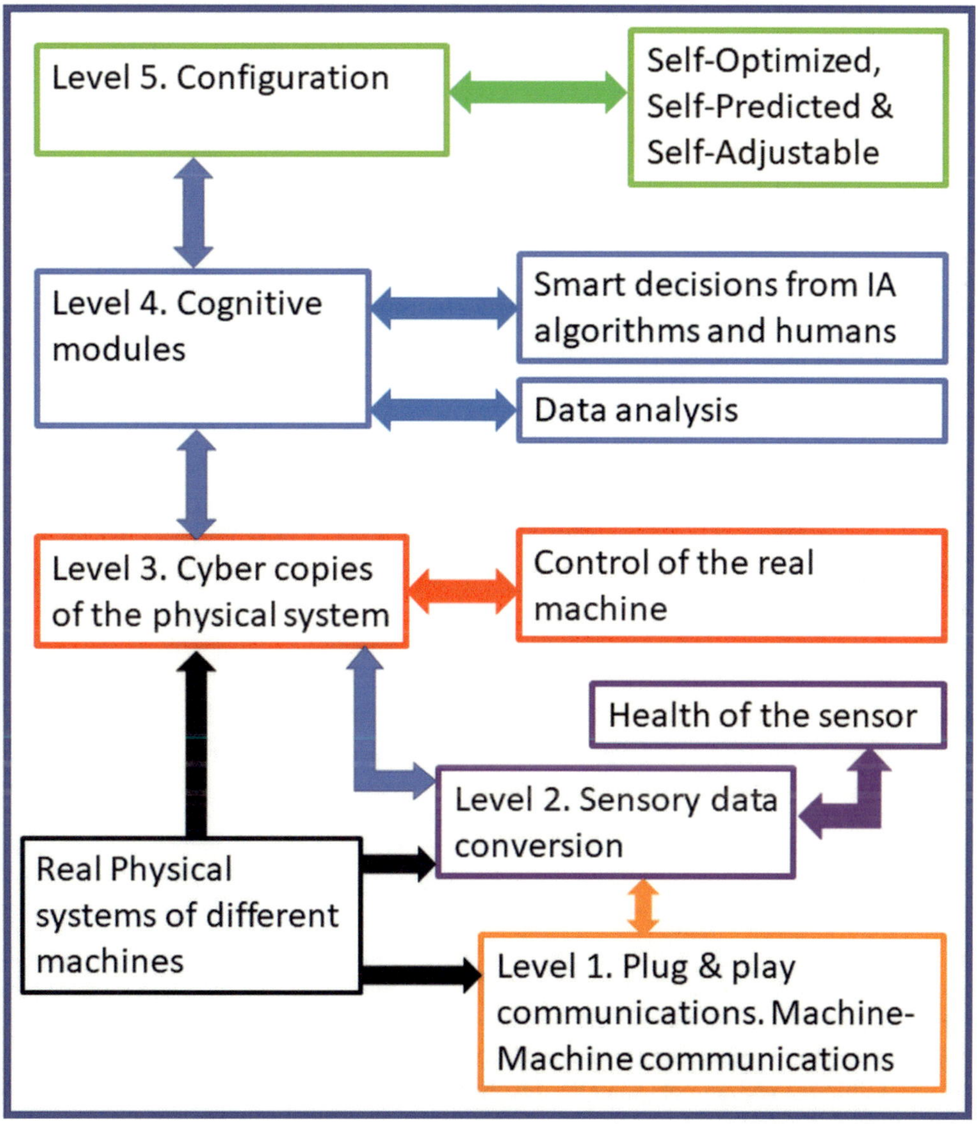

Figure 1. Cyber-physical architecture that includes five levels.

This communication can be used in the second level of conversion of data. In this level, several mechanisms can be used to convert the data into information in order to perform intelligent analysis of the data. Some mechanisms are developed for the forecast and management of the health of the machine, while others are used to analyze the degradation and prediction of performance and also to perform some correlation of multidimensional data.

The third level is created to generate virtual copies or cyber copies of the real physical systems. This level acts as central information where information is sent from each connected machine in a "virtual network of machines." This module will produce massive information or big data. This information should be stored for being analyzed to extract additional information that provides a better understanding of the status of the individual fleet machine.

These analyses provide the machines with a self-comparison capability, where the performance of a single machine can be compared and qualified among the fleet. On the other hand, the similarities between the performance of the machine and the previous assets (historical information) can be measured to predict the future behavior of the machinery. The historical information generated represents the memory that is accumulated over time of the data generated to identify any variation of the machine.

The fourth level is the level of intelligence and decision-making that provides an adequate presentation of the analytical information so that expert humans and algorithms can decide in the production process. This analytical information can be viewed remotely so that the operator can access the analyses and make the pertinent decisions using human-machine interfaces (HMI) for industry 4.0.

It also has the function of a collaborative diagnosis for maintenance processes, which can be easily determined due to the availability of comparative information and the status information of an individual machine.

The fifth level is the high level of the CPS where a configuration is made with the feedback from the cybernetic part to the physical part. This level performs supervisory control to make the machines self configured and self adaptive. It acts as the resilience control system (SCR) to apply the corresponding controls to the decisions made at the level of cognition. Its typical functions are self-configuration for resilience (ability to recover from a disturbance), automatic adjustment for variation, and self-optimization against disturbances.

This architecture proposes three levels of control for avoiding perturbations in real-time control applications: high-level, middle-level, and low-level.

The high-level control can be implemented in levels 4 and 5. These control modules can use cloud services, while the middle level of control can be the level 3 that can use a virtual model of the process and control the real-time system as a master-slave control, where the master device is the cyber version and the slave device is the real-time controller for low-level control. Thus, this system cannot generate delays and interruptions in communication among cloud services and low-level control.

4. Proposed actions for the CPS

Ten actions are proposed for minimizing the possibility of producing bad products in a manufacturing system that can be implemented in the cyber-physical system proposed; these actions are described below:

1. *The first proposed action* consists of producing intelligent connections by means of a plug and play and plug and produce concept, in a similar way as the rest of CPS architectures [12]. With this smart connection, it is expected to foresee the demand, the delivery time, and the number of warehouses or intermediaries in the sale through machine learning from the cognitive module. It is proposed to implement a customer relation management (CRM) system, which collects demand data, automates and personalizes sales processes, creates databases with information, and carries out a commercial follow-up. It is also proposed to implement an enterprise resource planning (ERP) system in the cognitive module that integrates the inventory among many types of modules. In this case, it is convenient to manage the manufacturing, human resources, sales, and supply chain modules.

2. *The second proposed action* involves converting the extracted data and solving the formula. The conversion aims to use the data obtained during the connection, first action, to measure the characteristics of critical problems and predict possible problems. Next, a state of each operation is created. The purpose is to raise awareness of the machines and analyze the data extracted in the connection. In this case, it is intended to solve the demand formula. For this, it is expected to calculate the required inventory by multiplying the demand (e.g., the weekly order) by the delivery time, by the number of locations required during the process, and by the level of demand variation (standard deviation of the demand).

3. *Third proposed action*. The systems can be autoconfigured based on the results of the cognition and of the criteria of priority that the company grants them. For example, if it is considering that the inventory level is high depending on the existing demand, an expert should skip an alert to warn during production and minimize tasks. In any production process, it is recommended to apply a manufacturing execution system (MES) that documents raw materials and finished products. MES allows controlling of resources, analyzing the production, and establishing data on the life cycle of the product. If this information fuses negative or contradicts the objectives of the company, this data could be decisive in decision-making. It is also proposed to establish a system of ERP. It involves making an inventory that collects information about the available machines and tools and connects both the company and the suppliers.

4. *The fourth proposed action* requires automating the control of the data of the virtual machines generated at the cyber level. Therefore, it is recommended to perform Key Performance Indicators (KPI), which monitors the data collected in each of the activities or processes that require a machine. The objective is to perform a self-comparison that predicts possible problems to try to improve the times recorded during the data collection. Thus, the

data interaction among all the machines can be analyzed at different times for predicting performances, efficiencies, and behaviors of each machine.

5. *Fifth proposed action.* The data analyzed by the cognitive level will do self-evaluations to the collected content and forecasting problems. Therefore it is required to establish algorithms that improve the efficiency in the value flow of the analyzed machines, that is, to develop the final VSM. This algorithm will depend on many factors, so it cannot be established generically. Each manufacturing company must attend to its own needs. For example, since cycle times are being monitored, alerts can be automated if these cycles are too long. These alerts can implement mechanisms for solving known problems. It is also recommended to apply a CMM that controls the flow of data, optimize the actual production times, automatically update the configuration of the machines, centralize the data, and store the data so that the cognitive modules can follow up.

6. *The sixth proposed action* is useful for the connection, conversion, cognition, and configuration levels to take care of the monitoring of the machines by means of self-detection about their behavior and their state. The conversion measures the collected data and the characteristics of the possible problems to offer the database a self-evaluation. Through the configuration, the machine can be reconfigured according to the requirements that have been established.

Thus, an alert system could be designed to notify the workers of possible faults of the machines. This warning can be directed to the expert worker by means of advanced HMI 4.0 that can help to visualize the situation and make decisions. Moreover, these warnings can detect in addition when some machines will provide a damaged part or can cause health hazards.

Therefore, the records of the cyber copies of the physical system will be useful to register existing anomalies describing the cause of the defect and a description of the hazard. Thus, a maintenance improvement process can be implemented [13], where the TPM is served with these actions.

1. *The seventh proposed action.* For the SMED, a link is needed to automate the data collection of the times, numbering each operation and detecting whether it has been considered an internal or external operation. With this data, it intends to offer cognition through a CPS. As the purpose of this methodology is to convert internal processes to external processes as much as possible, algorithms must be proposed to decide when this is possible. For the design of these algorithms, it is necessary to take note of all the real dangers in each machine, if an operator works on it while it is working. These dangers are numerous and depend on the device, so the design of the algorithm must be customized. There are activities that cannot be performed such as unlocking or changing a mold with a machine on because it is dangerous. Therefore, it is necessary to analyze the possibility of carrying out activities with the engine in motion that is currently carried out with a stopped machine without danger to the worker [14]. Augmented reality can be used in these types of activities, showing in real time the analysis of the machine.

2. *The eighth proposed action* (automation, CMM, and SCM) is to invest in automation systems and robotization processes. Digitization can help to increase load volumes without the need to increase time or strength. In addition, it minimizes movements. It is proposed to apply blockchain technology to transform the value chain of the production process. Another useful tool in automated systems is CMM measurements. These measurements allow efficiency during production, and this increases the quality of the products. It is about connecting the already configured machines to a blockchain network to record data and share them. The objective is to verify that everything works correctly. It is also recommended to manage the supply chain (SCM) by tracking the products, linking the company with the suppliers and consumers. This methodology is not incompatible with the ERP, but if ERP is established at a time of production, it will not be necessary to repeat it. For that reason, it is not applied as a recommendation in this action.

3. *The ninth proposed action* consists of implementing a statistical module based on the Six Sigma (σ) method in the cognitive level. Six Sigma is a statistical method implemented by engineer Bill Smith when he worked at the Motorola Company [15].

It is an efficient method to solve a problem with the aim of reducing the number of products with defects. In terms of measurements of positions from sensors, machines or robots could be used to optimize the process. This methodology will result in increased production quality, increased revenue, and increased customer satisfaction.

The objective is to reduce the defects produced reaching a maximum threshold of 3.4 Defects per Million of Opportunities (DPMO). Therefore, the use of Six Sigma can be considered as a process objective, where processes not only find fewer defects, but they do so with low variability and more consistently. Therefore, Six Sigma reduces the variation, so that the analyzed data can be delivered as expected reliable.

Moreover, some alarms can be implemented in the software to alert the maximum limits that the system can accept. Thus, the cyber level can generate data that can be processed, and prevention algorithms can be implemented to minimize defects in the measurements. Therefore, during the historical information, the analysis can prevent bad conditions of sensors, machines, or robots.

4. *The tenth proposed action*. This action is for helping to increase the level of adaptability in the cases of agile production. In order to react in a better way with the existing hardware of the plant that uses flexible manufacturing (i.e., robots, CNC, automated systems, among others), it is advisable that the cognitive level of the cyber-physical system may have modules of supervised learning, deep learning, and reinforcement learning. The combination of these three levels of learning could be useful to readapt the position of robotic arms to a new path that is required for producing a new task or product.

Supervised learning [16] is essential for the normal operation of the system. For example, a system that uses visual recognition can detect the patterns with the information that has been trained, but it is not robust to recognize objects with different conditions. However, deep learning tries to model high-level abstractions in data using architectures composed of multiple nonlinear transformations [17]. Therefore, deep learning could be used for detecting a variety of patterns and generating new adaptations.

Optimization algorithms are important for defining new optimal paths where optimal results can be validated by algorithms that contain intelligent observers or intelligent agents for a particular task. For example, this methodology can be used in the generation of new trajectories of a robotic arm that manipulates and grasps different objects adaptively.

On the other hand, reinforcement algorithms [18] can be used to generate rewards when an action is performed well. Here, expert operators can take also part in the decision-making process to validate or discredit an action. In this way, the overall process can be adapted to new situations, allowing the system to readapt the decision inside of the constraints of time that the task requires.

5. Conclusions

The new smart industry of industry 4.0 requires the integration of different technologies, methodologies, and cyber-physical systems in order to improve the level of efficiency and capability of adaptation in the manufacturing process.

High levels of digitalization are required to analyze the big data produced in the cyber level. Therefore, intelligent algorithms can use this information to minimize risks and efforts of the operator in human-machine collaboration and optimize the overall system, maximizing the value of the manufactured product.

The operations that initially give priority to an automated process are the start and stop services of the equipment, operations that detect defects, operations that add effort to the operator (such as loading or transport activities), and feeding operations. However, these concepts are not new because, in the eighteenth century, the Jidoka methodology emerged, which was initially called Autonomation. This method is very close to automation, and its objective was to provide intelligence to machines without the need for human supervision [19]. However, as the fourth revolution advances, companies have understood that the human being is always necessary and the one that brings logic to the cognition of systems. Even if the automation is full, the system needs a previous configuration [20].

An example of collaboration among humans and robots can be seen with the implementation of collaborative robots (cobots). These cobots can be used with humans to collaborate in tasks where the validation of the operator is necessary. Moreover, another type of collaboration is when humans can take decisions from the results of the cognitive modules where the operator has to validate the results or to provide a hybrid decision using the human intelligence and the results of the intelligent algorithms.

In this chapter, ten actions have been proposed that can be used in a typical cyber-physical architecture of five levels but oriented toward manufacturing with the objective to eliminate the seven *mudas* of the JIT. Many companies use ERP systems that can be benefited to integrate the actions described before and use the combination of other cognitive modules to analyze the information in multidimensional channels.

Moreover, the last action is suitable for agile manufacturing where some processes can be adapted to new strategy tasks by means of learning methods. Therefore, the time that requires the new adaptation will be an important constraint to consider for producing new optimal strategies. Thus, the system will require initial training with the objective to become an intelligent expert system with the validation of reinforcement algorithms.

Moreover, high levels of security are required to be implemented in the cyber-physical system in all levels of communication. This situation is very important because cloud services, servers, embedded systems, sensors, and programmable logic controllers (PLCs), among others, are used to communicate information, store the data, and analyze the data in the high levels of the CPS. Therefore, it is really important that all the information produced could be generated without noise and errors and the integrity of the data cannot be accessed for non-authorized people.

The collaboration among operators and machines must have high protocols of security in the decision-making process and reinforcement learning and avoid external attacks in the network of the CPS.

Finally, future developments will consist of analyzing new architectures for deep learning processes based on vision systems where the vision system in combination with intelligent agents can predict anomalies in the production and make corrections in real time.

Author details

Salvador Cobos Guzman*, Maria Dolores Cima Cabal, Francisco Machio Regidor and Lucia Alonso Virgos

*Address all correspondence to: salvador.cobos@unir.net

Universidad Internacional de la Rioja, Escuela Superior de Ingeniería y Tecnología (ESIT), Logroño, La Rioja, Spain

References

[1] Stachowiak A, Oleśków-Szłapka J. Agility capability maturity framework. Procedia Manufacturing. 2018;**17**:603-610. DOI: 10.1016/j.promfg.2018.10.102

[2] Pérez Rave J, La Rotta D, Sánchez K, Madera Y, Restrepo G, Rodríguez M, et al. Identificación y caracterización de mudas de transporte, procesos, movimientos y tiempos de espera en nueve pymes manufactureras incorporando la perspectiva del nivel operativo, Ingeniare. Revista Chilena de Ingeniería. 2011;**19**(3):396-408. DOI: 10.4067/s0718-33052011000300009

[3] Anderson D, Carmichael A. Essential Kanban Condensed. Seattle, Wash: Lean Kanban University Press; 2016

[4] Chhatrawat RS, Dixit A. Lean production system: A review. International Journal of Advance Engineering and Research Development. 2016;**3**(3):452-457. DOI: 10.21090/ijaerd.030375

[5] Lacerda A, Xambre A, Alvelos H. Applying value stream mapping to eliminate waste: A case study of an original equipment manufacturer for the automotive industry. International Journal of Production Research. 2015;**54**(6):1708-1720. DOI: 10.1080/00207543.2015.1055349

[6] Sacristán F. Reducción de los tiempos. Tecnica Industrial. 2019. Available from: http://www.tecnicaindustrial.es/tiadmin/numeros/66/77/a77.pdf [Accessed: June 10, 2019]

[7] Ahmadi A, Cherifi C, Cheutet V, Ouzrout Y. A review of CPS 5 components architecture for manufacturing based on standards. In: Proceedings of the 11th International Conference on Software, Knowledge, Information Management and Applications (SKIMA 2017); December 2017; Colombo, Sri Lanka. ffhal-01679977

[8] Bagheri B, Yang S, Kao H, Lee J. Cyber-physical systems architecture for self-aware machines in industry 4.0 environment. IFAC-Papersonline. 2015;**48**(3):1622-1627. DOI: 10.1016/j.ifacol.2015.06.318

[9] Morgan J, O'Donnell G. Enabling a ubiquitous and cloud manufacturing foundation with field-level service-oriented architecture. International Journal of Computer Integrated Manufacturing. 2017;**30**(4-5):442-458. DOI: 10.1080/0951192X.2015.1032355

[10] Chen J, Yang J, Zhou H, Xiang H, Zhu Z, Li Y, et al. CPS modeling of CNC machine tool work processes using an instruction-domain based approach. Engineering. 2015;**1**(2):247-260. DOI: 10.15302/j-eng-2015054

[11] Liu X, Cao J, Yang Y, Jiang S. CPS-based smart warehouse for industry 4.0: A survey of the underlying technologies. Computers. 2018;**7**(13):1-17. DOI: 10.3390/computers7010013

[12] Lee J, Bagheri B, Kao H. A cyber-physical systems architecture for industry 4.0-based manufacturing systems. Manufacturing Letters. 2015;**3**:18-23. DOI: 10.1016/j.mfglet.2014.12.001

[13] Sekine K, Arai K. TPM for the Lean Factory: Innovative Methods and Worksheets for Equipment Management. Portland, OR: Productivity Press; 2017

[14] Sabadka D, Molnar V, Fedorko G. The use of lean manufacturing techniques—SMED analysis to optimization of the production process. Advances in Science and Technology Research Journal. 2017;**11**(3):187-195. DOI: 10.12913/22998624/76067

[15] Remembering Bill Smith, Father of Six Sigma, iSixSigma. 2019. Available from: https://www.isixsigma.com/new-to-six-sigma/history/remembering-bill-smith-father-six-sigma/. [Accessed: June 10, 2019]

[16] Iosifidis A. Extreme learning machine based supervised subspace learning. Neurocomputing. 2015;**167**:158-164. DOI: 10.1016/j.neucom.2015.04.083

[17] Chen LC, Schwing AG, Yuille AL, Urtasun R. Learning deep structured models. In: Francis Bach and David Blei, (Eds.), Proceedings of the 32nd International Conference on International Conference on Machine Learning (ICML'15). Vol. 37. 2015. JMLR.org 1785-1794

[18] Vafashoar R, Meybodi M. Reinforcement learning in learning automata and cellular learning automata via multiple reinforcement signals. Knowledge-Based Systems. 2019;**169**:1-27. DOI: 10.1016/j.knosys.2019.01.021

[19] Baudin M. Working with Machines: The Nuts and Bolts of Lean Operations with Jidoka. New York: Productivity Press; 2007

[20] Åhlström P. Jidoka, wiley encyclopedia of management. Chichester, UK; 2015. DOI: 10.1002/9781118785317.weom100183

Chapter 5

Leveraging Internet-of-Things to Support Circular Economy Paradigm in Manufacturing Industry

Giuseppe Pacelli, Enrico Ferrera, Rosaria Rossini,
Ilaria Bosi and Claudio Pastrone

Additional information is available at the end of the chapter

http://dx.doi.org/10.5772/intechopen.80453

Abstract

Circular economy represents a fundamental alternative to the currently predominating linear economy model, while Industry 4.0 is a technological enabler to bring process innovation in the industrial domain. New economic models are needed in order to reduce material inputs and waste generation leveraging on ecodesign, recycling and reusing of products, new business models, and new technologies. Internet-of-Things and artificial intelligence can support the circular economy paradigm, through the development of a marketplace for connecting buyers and sellers of manufacturing services, raw materials and products toward building global supply chains. The core component of this marketplace is a novel, agent-based, brokering module that will apply both syntactic and semantic matching in terms of manufacturing capabilities, in order to find the best possible supplier to fulfill a request for a service, raw materials or products involved in the supply chain.

Keywords: circular economy, industrial symbiosis, Internet-of-Things, agent-based marketplace, artificial intelligence, waste management, trading

1. Introduction

Circular economy represents a fundamental alternative to the currently predominating linear economic model, based on the take-make-consume-dispose paradigm. The linear model has conducted to economic growth and welfare but ran its course. New economic models are needed in order to reduce material inputs and waste generation leveraging on ecodesign, recycling and reusing of products, new business models, and new technologies. Products and production systems need to be designed to be "circular": materials need to be efficiently processed and waste needs to be arranged and recycled.

Looking from a technological perspective, the circular economy generates needs in the field of manufacturing, processing, networking, identification, and recycling of materials and products. As Bányai discussed [1], the success of the operation of these functions is based on the optimal design and control. Most of the supply chain models include only the traditional procurement-production-distribution sub-processes used for converting raw materials to final products and deliver them to the wholesalers, retailers or directly to end users. In this case study is showed a complex green supply chain model in which sophisticated operation research heuristics has to be used to find the optimal solution in order to minimize the costs.

The main needs generated by circular economy are as follows:

- Advanced collection, sorting, and recycling technologies—apps, sensors, robots, etc.
- Efficient materials processing technologies—machine learning and artificial intelligence.
- Production technologies that support design for circularity—3D printing, disassembly, and repairability.
- Interactive platforms for enhanced connectivity—apps, websites, databases, and Internet-of-Things.

Industry 4.0 is fundamental to move from linear to circular economy model. Industry 4.0 refers to the introduction of new technologies, such as the Internet-of-Things, 3D printing, artificial intelligence, cloud computing, machine-to-machine communication, etc., in the traditional industrial domain in order to support the improvement of the efficiency and automation of the processes happening in the industry.

The introduction of circular economy generates new technological and nontechnological needs. The change in ownership and material management concepts, both at a consumer and at business level, generates a need for upscaling and acceleration of business concepts such as: products as a service, sharing platforms, peer-to-peer interactions, and industrial symbiosis [2, 3]. Industrial symbiosis approach focuses on the hidden value of waste resources within an industrial network which can be exploited through the cooperation [4].

Wasted resources are unused or reusable energy, water, materials, and residues of production processes. Industrial Symbiosis can help improve the overall efficiency of the industrial system: companies within an industrial symbiosis network establish beneficial business relationships among them based on these exchanges of resources. However, these relationships are not always easy to establish due to the difficulty in creating the network or lack of a communication platform for the network [5].

The vision of the circular economy is well suited to the concept expressed by the manifesto of the Physical Internet (PI) Initiative [6], a recent concept of breakthrough innovation aiming to improve by an order of magnitude; the economical, environmental, and societal efficiency and sustainability of the way physical objects are moved, deployed, realized, supplied, designed, and used. It attempts to achieve this scope by applying concepts from Internet data transfer to real-world shipping processes [7, 8].

Previous research has already proven that new inventory models enabled by and applied in PI could help reduce inventory levels, thanks to its high flexibility [9]. The vision of the physical Internet involves encapsulating goods in smart, ecofriendly, and modular containers ranging from the size of a maritime container to the size of a small box. That new inventory models enabled by and applied to PI could help reduce inventory levels, thanks to its high flexibility. These modular containers will be continuously monitored and routed, exploiting their digital interconnection through the Internet-of-Things.

Circular economy concepts, and more specifically industrial symbiosis, are vehiculating lot of research in the Internet-of-Things (IoT) domain. Indeed, waste characterization and identification can be considered as the most challenging steps in waste management, that is, the ones that kick off the subsequent valorization and exploitation. Once characterized and identified the waste, the IoT-domain literature proposes a huge number of infrastructures and platforms (e.g., see [10–12]), which helps to monitor features of interest, in real time or on historical base, communicating information only to subjects who are subscribed to specific topics of interest, just when they are available.

The aim of this chapter is to present how Internet-of-Things and artificial intelligence can support the circular economy paradigm, through the development of a marketplace for connecting buyers and sellers of manufacturing services, raw materials, and products toward building global supply chains. The core component of this marketplace is a novel, agent-based, brokering module that will apply both syntactic and semantic matching in terms of manufacturing capabilities, in order to find the best possible supplier to fulfill a request for a service, raw materials or products involved in the supply chain.

Section 2 provides the analysis of the background and state of the art of the multi-agent systems.

Section 3 of this chapter is dedicated to the description of a typical industrial symbiosis case study that the proposed solution intends to address.

Section 4 introduces the solution proposed by this work to support the circular economy, which integrates traditional IoT platforms with an agent-based approach to support a bidding system through the implementation of a marketplace.

Section 5 specifies the type of agents implemented in the marketplace, and finally Section 6 concludes the chapter addressing the future works to extend the platform.

2. Background and state of the art

Multi-agent systems (MAS) [13] have been widely studied in the literature, and some existing options have been evaluated before deciding to implement the marketplace. A multi-agent system is a computerized system composed of multiple interacting intelligent agents. Agent platforms are typically designed with a container paradigm, where all agents live in a well-defined agent container. Among the many platforms available on the Internet, a few subsets of candidates that might cover a good subset of composition features and requirements have been selected.

JIAC (Java-based Intelligent Agent Componentware) [14] is a Java-based agent architecture and framework that allows an easy development as well as the operation of large-scale, distributed applications, and services. The framework supports the design, implementation, and deployment of software agent systems. Then, the entire software development process is supported by JIAC. It also allows the possibility of reusing applications and services, also by modifying them during runtime. The strength points of JIAC are distribution, scalability, adaptability, and autonomy. In fact, JIAC V applications can be developed using extensive functionality that is provided in a library. This library has already-prepared services, components, and agents which can be integrated into an application. For example, individual agents are based on a component architecture which already provides the basic functionality for communication and process management. However, application-specific functionality can be provided by the developer and be interactively integrated.

SARL [15] is a general-purpose agent-oriented language. SARL focuses on providing the fundamental abstractions for dealing with interaction, concurrency, distribution, reactivity, decentralization, autonomy, and dynamic reconfiguration. These are high-level features considered so far as the major requirements for a good and practical implementation of modern complex software applications. SARL approach remains as generic as possible and highly extensible to easily integrate new concepts and features comparing to the variety of existing approaches and meta-models in the field of agent-oriented engineering and multi-agent systems. The language is platform and architecture independent.

JADE (Java Agent DEvelopment Framework) [16] is a software framework that eases the implementation of multi-agent systems through a middle-ware complies with the FIPA [17] specifications and a set of graphical tools that help the debugging and deployment process. Another helpful characteristic concerning JADE-based system is that it can also be distributed across machines with different OS. In a distributed scenario, the configuration of each system deployed can be controlled remotely thought a GUI. The configuration can be even changed at run time by moving agents from one machine to another, if required. Finally, JADE is completely implemented in Java language, and the minimal system requirement is the version 5 of JAVA.

SPADE (Smart Python multi-Agent Development Environment) [18] is a multiagent and organizations platform based on the XMPP/Jabber technology and written in the Python language. This environment offers features and facilities that simplifies the construction of MAS, examples are: the concepts of users (agents) and servers (platforms); an existing communication channel and an extensible communication protocol based on XML, just like FIPA-ACL [19]. Many other exist, but SPADE has been the first agent platform based on XMPP technology.

In **Table 1**, a comparative summary of the candidate agent platforms is reported, where respective pros and cons are addressed.

Platform	Pros	Cons	License
JIAC	Exploits Apache ActiveMQ as transport declarative agent definition through spring	Not FIPA compliant	Apache v2.0
SARL	Exploits ZeroMQ as transport	Not FIPA compliant	Apache v2.0
JADE	FIPA compliant	Supports only RMI transport. Few extensions for JMS/XMPP transports	LGPL
SPADE	FIPA compliant	Uses a custom XMPP server	LGPL

Table 1. Candidate agent platforms.

As it is possible to notice from the presented analysis, none of the platforms offers both a distributed approach and a general purpose implementation for the communication back end.

The platform proposed in this work, therefore, represents a unique example of a marketplace based on semi-automatic multi-agent collaboration.

3. Case study

Internet-of-Things can support the circular economy in many ways. A typical case study consists on the early and automated detection of scrap material levels and its management, also identifying a network of companies which take waste materials as raw material for their production process.

Whenever a desired level (e.g., 50, 75 or 90%) is detected by the scrap material storage area sensors, it will be annotated, reported, and recorded through IoT platform automatically. Getting accurate information about the scrap material storage area levels during the day will now be possible and the company producing the scrap material will stop occupying the staff time to manually detect scrap material levels at each storage area, with the system taking care of such procedure in each factory. Real time data will help optimize scrap material management procedures. Scrap material recycling companies will be either automatically contacted by the platform or they will receive continuous information on the current scrap material level. In response to such an interaction, they will be ready to put their offers to the system, once a new bidding process will be started by the company producing the scrap material. Such process can take place once the scrap material specification and the bidding specifications have been identified, with the company inserting the bid details in the platform. The system registers the new bid request and contacts the waste management companies for getting bids on scrap material. Companies contacted by the system typically belong to a selected subset of contractors. However, according to the company's policies, in some cases, the bid could be open to new contractors, for example, to find better offers fostering new collaborations. When the system contacts contractors for bidding, a deadline is provided on when the last bid will be accepted. Once the deadline is reached, the platform system will select the best (or n-best) offer according to the bidding process criteria (entered when the process started) and present

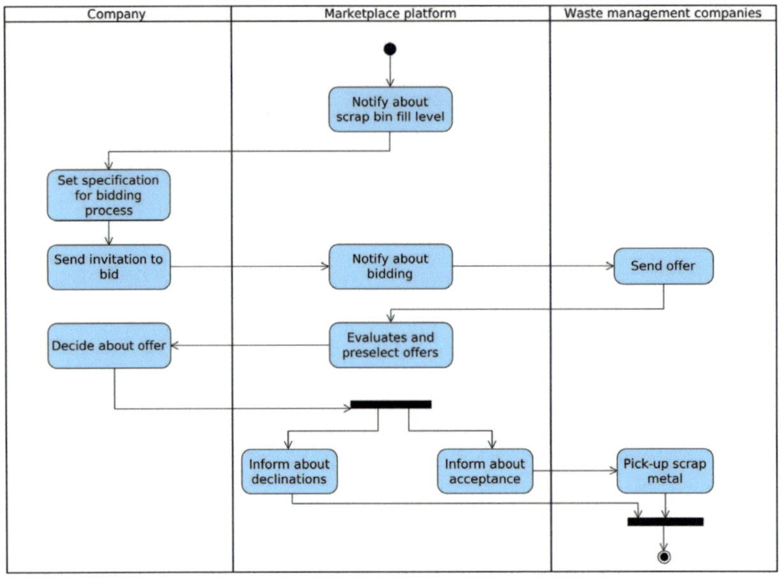

Figure 1. Scrap material collection and bidding process.

it to the relevant manager (this automation will reduce the time required for processing bids). Upon human approval, bidders will be automatically informed about the outcomes of their bids. The company will then arrange a pick-up date and time in agreement with the selected scrap material recycling company; such an arrangement will be automatized to the greatest extent (see **Figure 1**).

The criteria to award the contract will be price, company reputation, and the data agreement for the collection of the material. The literature presents several well-known algorithms that can be leveraged for automatic negotiations, such as contract net protocol [20], while many protocols are nowadays considered de facto standards for supporting secure peer-to-peer communication in IoT (such as, XMPP, MQTT, CoAP, and AMQP)

4. The bidding platform a.k.a. marketplace

This section introduces the solution proposed to support the circular economy in the use case presented in Section 3. The proposed platform integrates traditional IoT platforms with an agent-based approach to support a bidding system through the implementation of a marketplace. The marketplace is a fully distributed multi-agent system designed to support Industry 4.0 exchanges between involved stakeholders. It is particularly aimed to supporting automatic supply chain formation and negotiation of goods/data exchanges. The marketplace exploits a microservice architecture, based on Docker, and relies upon a scalable messaging infrastructure. Trust and security are granted in every negotiation step undertaken by automated agents on behalf of involved stakeholders.

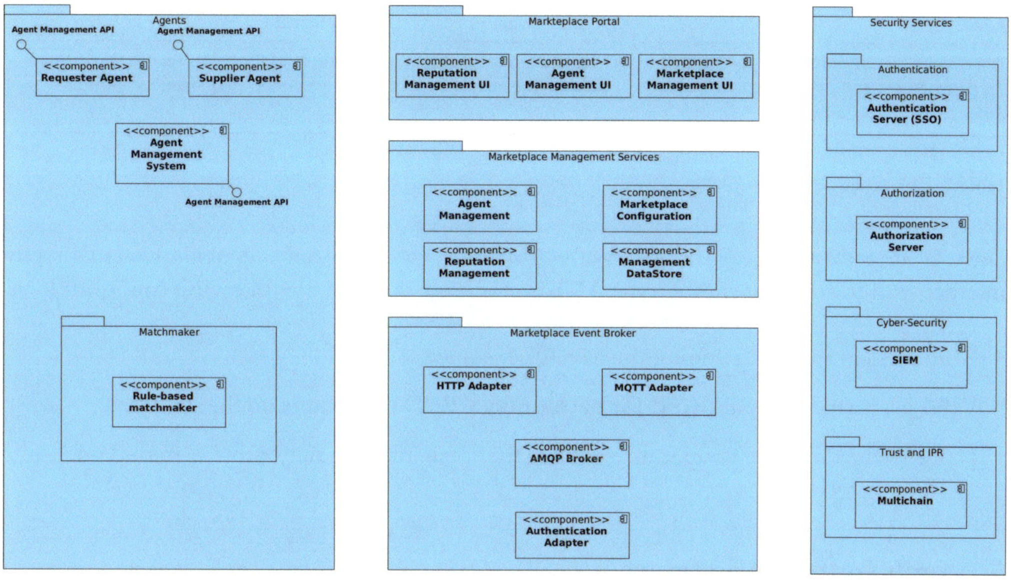

Figure 2. Marketplace components overview.

The marketplace includes the following elements: (a) a marketplace management portal; (b) a marketplace core based on Docker; and (c) a set of "default" agent implementations ready to be adopted by interested stakeholders.

The main building blocks of the marketplace are displayed in **Figure 2**.

4.1. Agent management system

According to FIPA specifications [21], an Agent Management System (AMS) is a mandatory component of every agent platform and only one AMS should exist in every platform. It offers the White Pages service to other agents on the platform by maintaining a directory of the agent identifiers currently active on the platform. Prior to any operation, every agent should register to the AMS to get a valid agent identifier, which will be unique within the agent platform.

4.2. White Pages service

A White Pages service is a mandatory component of any MAS system. It is required to locate and name agents on the system, making it possible for one agent to connect with one another. In the current implementation of the Agent Management System, the agent identifiers are stored in a MySQL database. MySQL has been chosen because it offers relevant features for the project such as on-demand scalability, high availability, and reliability. Other agent platforms, like SPADE, use MySQL as well for the database underlying the White Pages service.

Agents are stored according to the schema depicted in (**Table 2**):

Agent_id	Agent_owner	Agent_role
The unique identifier for the agent. It is the primary key for the table, having the constraint of being unique.	The name of the company owning the agent.	The role of the agent on the marketplace can be either 'Requester' or 'Supplier'.

Table 2. Table schema.

Since the directory service is offered by the AMS, the latter is the only agent allowed to directly interact with the database. When the AMS is executed, it needs the following configurations:

- IP address of the host running the MySQL instance.
- Valid username and password for performing CRUD operations on the database.
- Name of the database to operate on, since the same instance may run different databases.

5. Agents in the marketplace

There are two main categories of agents that can be defined a priori, depending on the kind of services provided:

- Marketplace agents (MA)
- Stakeholder agents (SA)

The first category groups all the agents providing services that are crucial for the marketplace to operate. The second category, instead, groups agents developed and deployed by the marketplace stakeholders to take part in chain formation rounds.

From an implementation point of view, they are very similar and share a large set of features, especially the communication protocols used for the interaction with stakeholders and other agents.

The communication protocols used by the MA for the interaction with the stakeholder on left side and with other agents on the marketplace on the right side are shown in **Figure 3**. Any agent can be controlled by the stakeholder via RESTful APIs which will be listed and explained in the following sections. Agents on the marketplace can exchange messages using Advanced Message Queuing Protocol (AMQP) as transport layer.

Stakeholder agents are instead deployed at the stakeholder's premises, and their purpose is to fulfill the stakeholder's interests. Stakeholder agents can be further divided into two different categories: requester and supplier. In the following, the reference implementations for the two different kinds of stakeholder agents will be described. However, both kinds of agents share a set of features allowing them to respond in different ways to the same kind of solicitations (such as input from GUI, messages from other agents) according to the state of their current behavior (when the solicitation is received).

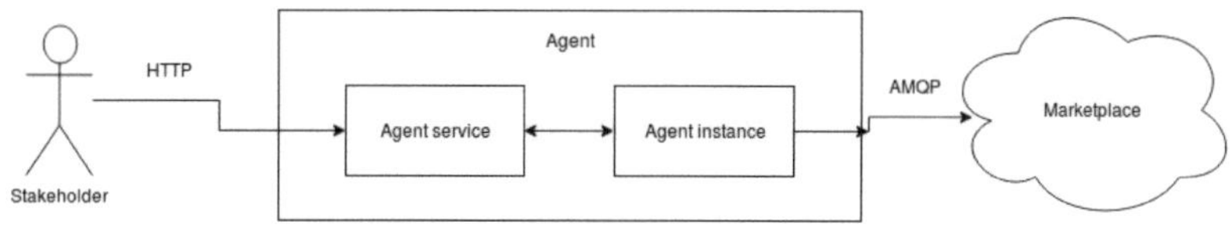

Figure 3. Simplified agents' communication logic.

Currently, agents support two different behaviors:

1. Authentication behavior: This is triggered upon the each agent activation. It consists of all the procedures that are necessary for the agent to correctly operate on the marketplace.

2. Negotiation behavior: This is the behavior each agent needs to support in order to be able to participate in marketplace negotiations. It is designed in a way that allows different negotiation protocols to be adopted.

The authentication behavior is the first one engaged by any agent since:

1. It takes care of getting the agent identifier from the AMS.

2. It authenticates the agent with the messaging broker, allowing secure message exchange over SSL.

Both these phases are mandatory for any agent to operate on the marketplace. After the authentication process, the negotiation behavior comes to play. It allows the agent to perform transactions and interactions with the other agents on the marketplace. Its behavior is constrained to the negotiation protocol that is being used.

Agents might or might not support a certain level of "intelligence" in their decisions regarding certain negotiation protocols. For example, an agent might have the capability of self-evaluation for the received offers, whereas another agent might not have such capability and, therefore, it will involve an external agent (providing such intelligence) during the decision process. An agent having such level of intelligence will be described as "smart," while it will be described as "dumb" in the absence of that.

5.1. Requester agent

The requester agent is the agent used by a factory to request the execution of an existing supply chain or to initiate a new supply chain. Due to the dynamics of exchanges pursued in a factory-oriented marketplace, there is no actual distinction between the two processes, that is, for any supply need a new chain is formed, and a new execution of the chain is triggered. The requester agent may act according to several negotiation protocols, which can possibly be supported by only a subset of the agents active on a specific marketplace instance. In the proposed implementation, it exists a baseline protocol, which must be supported by any agent, called CONTRACT-NET. In such protocol, a requester agent plays the role of "Initiator."

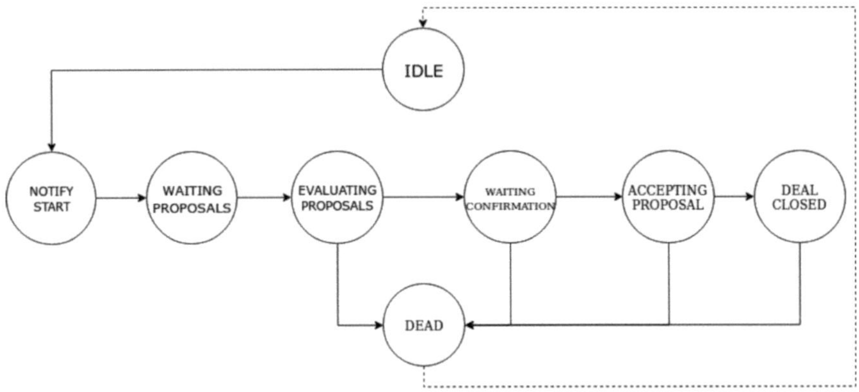

Figure 4. Requester states.

As described by Shoham [22], any agent can pass through a different set of states; the state of an agent consists of components such as beliefs, decisions, capabilities, and obligations. In the CONTRACT-NET, protocol requester agents go through different sets of states, performing different actions upon the receipt of a message (either from the UI or from other agents) according to their current state. In **Figure 4**, the states for requester agent are shown with arrows indicating the allowed transitions between one state and another.

5.1.1. *Requester agent negotiation protocol*

One of our factory-oriented marketplace strengths is the tight connection existing between intra-factory and inter-factory components. This connection allows the agent on the marketplace to operate in the optimum way to fulfill the stakeholder's needs.

Concerning the requester agent, such connection is the starting point of a new negotiation protocol: the IIMS notifies, through the learning agent (an agent available at the stakeholder's premises, in charge of intra-inter factory communication), that a certain good/service is needed. Upon receipt of such notification, the requester agent can react accordingly, using the most suitable protocol to fulfill the request. A possible interaction is here described from the requester agent point of view:

1. Marketplace system automatically notifies the stakeholder about the scrap metal bin fill level, through a notification from the learning agent (which receives the data from the sensors installed in the factory premises) to the requester agent.
2. Requester agent has the logic for starting a new bidding process through default parameters (set by the stakeholder).
3. Requester agent starts the bidding process, looking for the list of the supplier agents currently available on the marketplace that are capable in replying to such offer.
4. Offers come to requester agent, which can either evaluate them locally (smart agent behavior) or send them out for evaluation toward an external agent.

5. Once the deadline for receiving proposals has expired, requester agent notifies the UI about the best one(s) that have been received.
6. Stakeholder can now choose the preferred offer, select it, and notify the requester agent that can close the deal with the corresponding supplier agent.

The same input that is needed for a bidding process to start, that is, the fill level, can be used by the requester agent to update the supplier agents interested in receiving such information in order to be ready when a new bidding process will take place. The latest described scenario is here described from the requester agent's point of view:

1. The fill level information is received from the learning agent.
2. The requester agent forwards such information, in the marketplace, only to the supplier agent interested in receiving it.

5.2. Supplier agent

The supplier agent is the counterpart of the requester agent on marketplace. It is usually adopted by actual suppliers to respond to supply requests coming from other stakeholders in the marketplace. Factories transforming goods typically employ at least one requester agent to get prime goods and one supplier agent to sell intermediate products to other factories.

In the CONTRACT-NET protocol, supplier agents go through different sets of states, performing different actions upon receipt of a message (either from the UI or from other agents) according to their current state. In **Figure 5**, the states for supplier agent are shown with arrows indicating the allowed transitions between one state and another.

5.2.1. Supplier agent negotiation protocol

A typical supplier agent is registered on the marketplace remaining "silent" while waiting for a new call for proposal, issued by a requester agent. Once this message arrives, the supplier agent:

1. Evaluates the call for proposal message according to the policies set by the agent owner.
2. If the proposal is compliant with such policies, an offer is sent from the supplier agent, otherwise it goes back to the "silent" (IDLE) state.
3. Once the offer has been sent, the supplier agent waits for the requester's decision.
4. In the end, if the requester agent has accepted the proposal, a notification from the supplier agent is sent to the UI for notifying the stakeholder and waiting for his decision.
5. The final message for acceptance is sent back to the requester agent in case the stakeholders accept the deal, otherwise a reject message will be sent.

As already mentioned for the requester agents, the tight connection between intra-factory and inter-factory systems is a strength for the marketplace. For the supplier agent, this connection is exploited to dynamically adapt to the market needs.

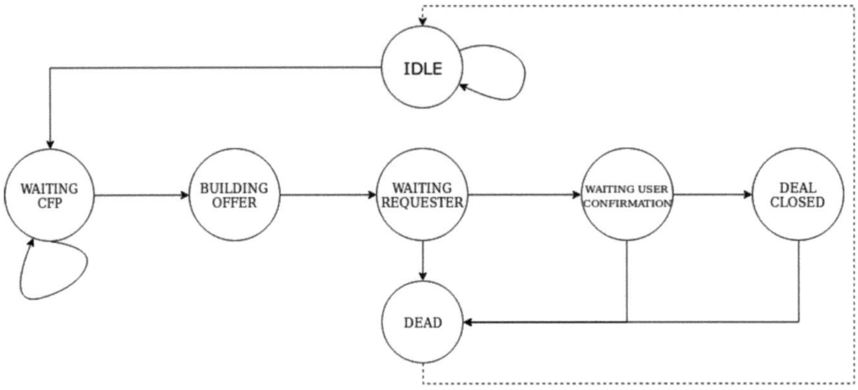

Figure 5. Supplier states.

For the supplier agent, this connection is here briefly described from the supplier agent point of view:

1. The supplier agent subscribes, via AMQP, to the queues where the fill level notification related to a good of a certain company will be received.
2. When such information arrives, the supplier agent can store it to exploit it for predictions of future fill levels.

The same behavior can be used for predicting the prices of a certain good/service. The supplier agent can, in fact, store information about the past bidding process and apply deep learning techniques to have a prediction about the possible future values of interest, that is, what will be the selling price for a certain good at a certain point of time in the future.

6. Conclusions and future works

The proposed marketplace platform shows an approach that enables the paradigm switch from the linear economy model to the circular one. The circular economy paradigm is based upon reuse, reparation, refurbishment, and recycling of existing materials and products and, as shown in the presented case study, these activities can be facilitated through the proposed marketplace approach. Particularly, a company can order the raw materials exploiting a certain supply chain available on the marketplace, process them, and sell the residual production waste through another dedicated supply chain available on the marketplace. On the other hand, a recycling company can buy (or simply collect) the offered waste, properly process it, and sell it again as raw material.

What is important to notice is that, even though existing collaboration models can be exploited on the marketplace, the proposed implementation has been designed to foster new collaboration models within and across the supply chain, making it easier for new actors to participate and collaborate with the existing ones. In order to increase the overall market competitivity while trying to maximize the participant profits, new protocol interactions based on different kinds of auctions are being studied. For instance, Vickrey auction [23] is being investigated, since it has been proved to be efficient, under certain circumstances, in maximizing the expected utility of each bidder.

One of the next important issues to be carefully addressed concerns the trust. Platforms like eBay imply the complete trust of strangers who take money before sending any wished goods. The reputation of sellers is built according to feedback provided by previous buyers and often is very hard for newcomers to sell items on the marketplace because they have to build up a good reputation. With this trust system is very easy to fake reviews in such a way to increase the reputation. Nowadays, Distributed Ledger Technologies (DLT), such as blockchain, are analyzed to be applied in many different application domains. Marketplaces can benefit from the power of DLT since it is built to create trust and transparency. In order to enhance the proposed marketplace approach, the authors are planning to leverage on DLT. DLT have received more and more attention in recent years as innovative methods to store and update data within and among organizations. Multiple copies of the ledger are held by different parties, with data added by consensus and without the need for a third party. This means that DLT are able to offer an immutable record, which means that data added to the ledger is in theory unchangeable, secure, and preserved for the life of the ledger, with the agreement of all participants. Additions to the ledger or changes to the governing structure are decided on a consensus basis by multiple participants. So, these systems provide a transparent and verifiable record of transactions. As a result, DLT can provide gains in trust among bidding participants.

Acknowledgements

This work was supported by the European Union's Horizon 2020 research and innovation program through the COMPOSITION project (Grant No. 723145).

Author details

Giuseppe Pacelli[1], Enrico Ferrera[1*], Rosaria Rossini[2], Ilaria Bosi[1] and Claudio Pastrone[1]

*Address all correspondence to: ferrera@ismb.it

1 Istituto Superiore Mario Boella—ISMB, Turin, Italy

2 Leading Innovation & Knowledge for Society—LINKS, Turin, Italy

References

[1] Bányai T. Optimisation of a multi-product green supply chain model with harmony search. In: DAAAM International Scientific Book. 2011. pp. 15-31

[2] Homrich AS, Galvão G, Abadia LG, Carvalho MM. The circular economy umbrella: Trends and gaps on integrating pathways. Journal of Cleaner Production. 2018;**175**: 525-543

[3] Ferrera E et al. Toward industry 4.0: Efficient and sustainable manufacturing leveraging MAESTRI total efficiency framework. In: Campana G, Howlett R, Setchi R, Cimatti B, editors. Sustainable Design and Manufacturing 2017. SDM 2017. Smart Innovation, Systems and Technologies. Vol. 68. Cham: Springer; 2017

[4] Holgado M, Morgan D, Evans S. Exploring the scope of industrial symbiosis: Implications for practitioners. In: Sustainable Design and Manufacturing. Cham: Springer; 2016. pp. 169-178

[5] Sakr D, Baas L, El-Haggar S, Huisingh D. Critical success and limiting factors for eco-industrial parks: Global trends and Egyptian context. Journal of Cleaner Production. 2011;**19**(11):1158-1169

[6] Montreuil B. Physical Internet Manifesto, Version 1.11. 1. CIRRELT Interuniversity Research Center on Enterprise Networks, Logistics and Transportation; 2012. Specific reference: Montreuil B. Toward a Physical Internet: Meeting the global logistics sustainability grand challenge. Logistics Research. 2011;**3**(2-3):71-87

[7] Yang Y, Pan S, Ballot E. Mitigating supply chain disruptions through interconnected logistics services in the physical internet. International Journal of Production Research. 2017;**55**(14):3970-3983

[8] Pan S, Ballot E, Huang GQ, Montreuil B. Physical Internet and Interconnected Logistics Services: Research and Applications. 2017;2603-2609

[9] Pan S, Nigrelli M, Ballot E, Sarraj R, Yang Y. Perspectives of inventory control models in the physical internet: A simulation study. Computers & Industrial Engineering. 2015;**84**:122-132

[10] Bonino D. ALMANAC: Internet of things for smart cities. In: 3rd International Conference on Future Internet of Things and Cloud, Rome; 2015. pp. 309-316

[11] Kamienski C, Jentsch M, Eisenhauer M, Kiljander J, Ferrera E, Rosengren P, Sadok D. Application development for the Internet of Things: A context-aware mixed criticality systems development platform. Computer Communications. 2017;**104**:1-16

[12] Brizzi P, Lotito A, Ferrera E, Conzon D, Tomasi R, Spirito M. Enhancing traceability and industrial process automation through the VIRTUS middleware. In: Proceedings of the Middleware 11 Industry Track Workshop; ACM; 2011. p. 2

[13] Wikipedia. Multi-Agent System. https://en.wikipedia.org/wiki/Multi-agent_system [Accessed: July 20, 2018]

[14] Lützenberger M, Küster T, Konnerth T, Thiele A, Masuch N, Hessler A, Keiser J, Burkhardt M, Kaiser S, Albayrak S. JIAC V—A MAS framework for industrial applications. In: 12th International Conference on Autonomous Agents and Multiagent Systems 2013; AAMAS; 2013

[15] SARL. General-Purpose Agent-Oriented Programming Language. http://www.sarl.io/ [Accessed: July 20, 2018]

[16] Bellifemine F, Poggi A, Rimassa G. JADE–A FIPA-compliant agent framework. In: Proceedings of PAAM. Vol. 99. No. 97-108; 1999

[17] FIPA. The Foundation for Intelligent Physical Agents. http://www.fipa.org/ [Accessed: July 20, 2018]

[18] Python Software Foundation. SPADE: Smart Python Multi-Agent Development Environment. https://pypi.python.org/pypi/SPADE [Accessed: July 20, 2018]

[19] FIPA. Agent Communication Language Specifications. http://www.fipa.org/repository/aclspecs.html [Accessed: July 20, 2018]

[20] Smith RG. The contract net protocol: High-level communication and control in a distributed problem solver. IEEE Transactions on Computers. 1980;**C-29**(12):1104-1113

[21] FIPA. FIPA Agent Management Specification http://www.fipa.org/specs/fipa00023/SC00023K.pdf [Accessed: July 20, 2018]

[22] Shoham Y. Agent-oriented programming. Artificial Intelligence. 1993;**60**(1):51-92

[23] Wikipedia. Vickrey Auction. https://en.wikipedia.org/wiki/Vickrey_auction [Accessed: July 20, 2018]

Chapter 6

Parallel Genetic Algorithms with GPU Computing

John Runwei Cheng and Mitsuo Gen

Additional information is available at the end of the chapter

http://dx.doi.org/10.5772/intechopen.89152

Abstract

Genetic algorithms (GAs) are powerful solutions to optimization problems arising from manufacturing and logistic fields. It helps to find better solutions for complex and difficult cases, which are hard to be solved by using strict optimization methods. Accelerating parallel GAs with GPU computing have received significant attention from both practitioners and researchers, ever since the emergence of GPU-CPU heterogeneous architectures. Designing a parallel algorithm on GPU is different fundamentally from designing one on CPU. On CPU architecture, typically data or tasks are distributed across tens of threads or processes, while on GPU architecture, more than hundreds of thousands of threads run. In order to fully utilize the computing power of GPUs, the design approaches and implementation strategies of parallel GAs should be re-probed. In the chapter, a concise overview of parallel GAs on GPU is given from the perspective of GPU architecture. The concept of parallelism granularity is redefined, the aspect of data layout is discussed on how it will affect the kernel performance, and the hierarchy of threads is examined on how threads are organized in the grid and blocks to expose sufficient parallelism to GPU. Some future research is discussed. A hybrid parallel model, based on the feature of GPU architecture, is suggested to build up efficient parallel GAs for hyper-scale problems.

Keywords: GPU acceleration, parallel genetic algorithms, granularity of parallelism of parallel genetic algorithms on GPU, GPU-CPU heterogeneous systems

1. Introduction

GAs are stochastic and global stochastic search methods, which combine two major search strategies: exploiting better solutions and exploring the global search space. GAs have been successfully applied to many optimization problems in different disciplines that are difficult to solve by conventional mathematical programming methods [1]. GAs, as a population-based search method, present great potential in performance improvement by parallel computation. The earliest attempt to parallelize GAs can be traced back to nearly 30 years ago. The research on parallel GAs can be divided into two major stages as the computing architectures evolve: the stage of parallel GAs on CPU architecture versus on GPU architecture.

On CPU architecture, you may run a parallel program over tens of threads or processes, while on GPU architecture, you can run it with more than hundreds of thousands of massive threads simultaneously, which leads to a fundamental paradigm shift in parallel programming. The basic idea to parallel GAs on CPU is to distribute computations among either multiple processors or multiple computers. Each instance of parallel processes on a processor is essentially a sequential GA, viewing from the perspective of algorithm implementation. The approach of parallel GAs on CPU is not suitable for massive GPU thread computing architecture.

Since the emergence of CPU + GPU heterogeneous architecture, how to accelerate GAs on the architectures has received significant attentions from both practitioners and researchers. GPU computing makes the research of parallel GAs truly enter into the world of high-performance computing (HPC) and demonstrates a great potential to many research disciplines and industrial worlds that can benefit from the power of GPU-accelerated stochastic and global search to explore large and complex search spaces for better solutions.

Most works on parallel GAs on GPU are published before 2013 on very earlier generation of GPU architectures, and most researches adopt a naive approach: let each GPU thread do a sequential GA, exactly the same way parallel GAs do on CPU. Even with the naive approach, you can still speed up GA computation greatly. Since massive threads are available on GPU, the way of designing a parallel program on GPU is quite different with the way of designing one on CPU. The distinguishing features of parallel programming on GPU over CPU are that CUDA platform exposes GPU memory model and execution model to us to enable us to have more control over massive threads in order to exert the computing power of GPUs.

The intention of the chapter is to give a concise review on accelerating GAs on GPU: how to organize the massive threads to access memories and how to harmonize massive threads to work efficiently. Fermi architecture, the world's first complete GPU computing architecture released in 2010, makes GPU computing truly applicable in industrial worlds. This is the reason why the review on the selected works after 2010. The performance issue is not touched, because since then, both GPU hardware architectures and CUDA platform have had significant progress.

This chapter is organized as follows. Section 2 briefly describes the basic concept of GA and parallel models on CPU. Section 3 discusses major issues on how to accelerate parallel GAs on GPU: the granularity of parallelism is redefined for parallel GAs on GPU, the aspect of data layout is examined to show how it will affect the kernel design to maximize memory bandwidth, and the hierarchy of thread execution is investigated to explain how to organize threads in grid and blocks to expose sufficient parallelism to GPU. A concise overview is given in Section 4 on selective works of parallel GAs on GPU. How to implement three parallel models of GAs on GPUs is explained through several famous combinatorial optimization problems to show that the efficient implementations of parallel GAs on GPUs share the common principles of kernel optimization. A case study of railway scheduling problem is described in Section 5 to demonstrate how to accelerate multiple objective parallel GAs with GPU. A brief discussion of future research is given in Section 6, focusing on how to build up efficient parallel GAs on GPUs for hyper-scale computing. The chapter ends with a brief conclusion in Section 7.

2. Parallel GAs on CPU

This section introduces briefly the basic concepts of GAs and basic models of parallel GAs on CPU.

2.1. General structure of genetic algorithms

GAs are efficient and stochastic search methods based on principles of natural selection and genetics. Many optimization problems from the scientific and industrial research are very complex in nature and quite hard to solve by conventional optimization techniques. Since the 1960s, there has been an increasing interest in imitating living beings to solve such kinds of hard optimization problems. Simulating the natural evolutionary process of human beings results in stochastic optimization techniques called genetic algorithms or evolutionary algorithms, which can often outperform conventional optimization methods when applied to difficult real-world problems [2].

The canonical form of GAs is described by Goldberg [3]. The calculation of GAs begins with a set of individuals, called a *population*. Each individual represents a solution to the problem you want to solve, which is also called *chromosome*. GAs perform the stochastic search through a loop of iterations, and each iteration is called a *generation*. New solutions are created at each generation by either merging two solutions using a *crossover* operator or modifying a solution using a *mutation* operator. The new solution is called *offspring* and the solutions used to create them called *parents*. During each generation, each solution in the population is evaluated, using some measures of *fitness*. A set of new population for the next generation is then formed by a *selection* operator. The solutions with better fitness values have much higher possibilities of being selected to enter the next generation. The population evolves over several generations, converging to the better solutions which are hopefully close to the optimal solution. The pseudo code of the general structure of GA is described as follows (**Figure 1**).

GAs, as a metaheuristic, have many advantages over the conventional single-point search methods. Requiring not good mathematical properties about the problem as the conventional method does, GAs can handle much complex problems with any kind of objective functions and constraints, linear or nonlinear, on discrete, continuous, or mixed search spaces, and provide us a great flexibility to hybridize with domain-dependent heuristics to make the random search more powerful and effective. The ergodicity of genetic operators makes GAs very effective at performing global search. GAs often outperform over conventional methods on most types of combinatorial optimization problems because they do not make any assumption about the underlying fitness landscape [4].

```
Procedure: Genetic Algorithms
begin
  t ← 0;
  initialize P(t);
  evaluate P(t);
  while (not termination condition) do
    recombine P(t) to yield C(t);
    evaluate C(t);
    select P(t + 1) from P(t) and C(t);
    t ← t + 1;
  end
end
```

Figure 1. The general structure of GA.

2.2. Multi-objective genetic algorithms

Multi-objective optimization deals with problems of seeking solutions over a set of possible choices to optimize multiple criteria. Almost every important real-world decision problem involves multiple and conflicting objectives that need to be tackled while respecting various constraints, leading to overwhelming complexity for conventional optimization algorithms. Genetic algorithms have been receiving considerable attention as a novel approach to multi-objective optimizations since the early 1980s, resulting in a fresh body of research and applications, called multi-objective GAs.

The primary difference of multi-objective GAs from single-objective GAs is how to determine the fitness value of individuals according to multiple objectives. For multi-objective problems, we cannot find a point superior than all other points in the criterion space, as illustrated in **Figure 2**, such points called non-dominated, or non-inferior, or Pareto point, named after an Italian economist, who introduced the concept of Pareto efficiency in the field of microeconomics. A point is called Pareto optimal if none of the objectives can be improved in value without degrading some of the other objectives' values.

How to maintain a set of non-dominated individuals during evolutionary processes is a special issue for multi-objective GAs. Pareto points are identified at each generation and used to calculate fitness values of all points or rank all other points. No mechanism in canonical GAs is provided to guarantee Pareto points in current generation be selected into next generation. In other words, some Pareto points may get lost during the evolutionary process. To avoid such sampling errors, a special mechanism for preserving Pareto solutions is added to the basic structure of genetic algorithms. In each generation, a set of Pareto points, kept in a separate pool, is updated by deleting all dominated solutions and adding all newly generated Pareto solutions [2].

http://dx.doi.org/10.5772/intechopen.89152

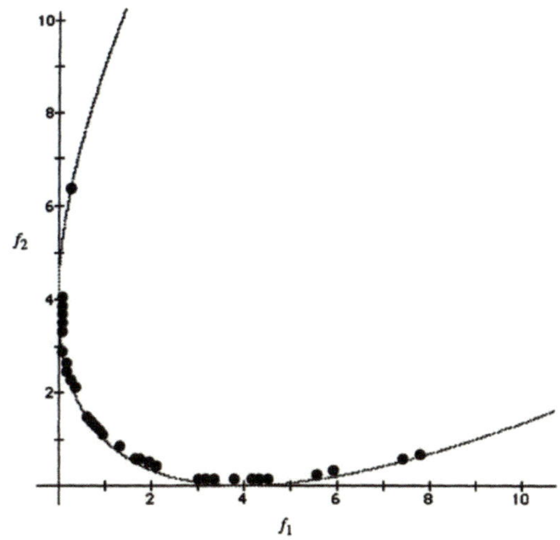

Figure 2. The Pareto frontier.

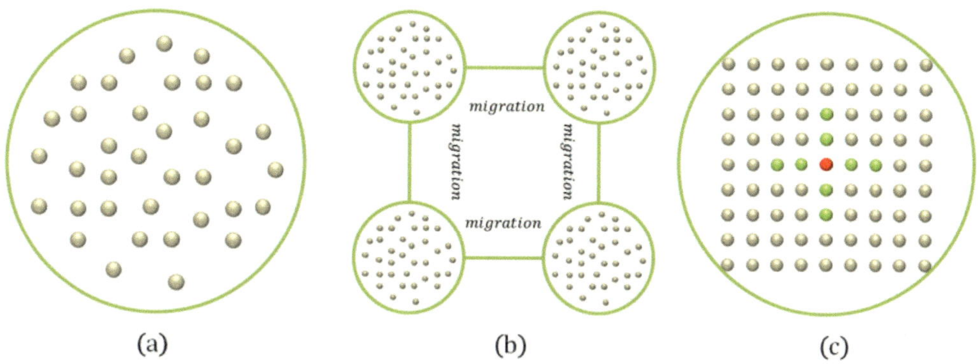

Figure 3. Models of parallel GAs. (a) Master-Slave model, (b) Island model, and (c) Cellular model.

2.3. Parallel genetic algorithms

Because GAs conduct the stochastic search over a set of solutions, there is a great potential for speeding up the evolutionary process of GAs: distributing the computational load among multiple processors through a data parallel approach. The motivation to parallelize GAs is twofold: to speed up the computation of GAs when solving large and complex problems and to improve the quality of solutions by exploiting distributed populations. There are three basic approaches of parallel GAs on CPU as shown in **Figure 3**:

1. Master-slave model
2. Island model
3. Cellular model

The basic idea behind all above algorithms is a divide-and-conquer approach: dividing the task into chunks and solving the chunks simultaneously using multiple processors [5]. In the *master-slave* model, there is a single population, and the evaluation of fitness is distributed among several processors. In the *cellular* model, there is a spatially structured population, and selection and mating are restricted to a small neighborhood, but neighborhoods overlap, permitting some interaction among all the individuals. This model is suited for massively parallel computers. In the *island* model, the population is divided into multiple sets, each called an *island*. In an island, individuals mate freely, while mating is prohibited across islands. A new operator is then introduced into the island model, the migration operator, exchanging a small portion of population among islands periodically in a predefined way in order to bring new genetic materials into each island.

From the perspective of mating mechanisms, these models can be classified into two major categories: mating globally and mating locally. In the master-slave model, individuals are permitted to mate freely within the entire population. In the cellular model, the population is imposed with geographical structure, and individuals are permitted to mate only with its close neighbors. The island model is in between the master-slave and cellular model: free within a given island but restricted among islands.

From the perspective of population, these models can be classified into two categories: the single population and the multiple populations. The master-slave and cellular models belong to the single population, and the island model belongs to the multiple populations.

Conceptually, all above models keep multiple instances of GAs, and each instance on a processor or a computer is a sequential program in nature. A communication mechanism among processors or computers is required in order to evolve a better solution from the overall populations, implemented either in a synchronizing way or in an asynchronizing way.

3. Major issues of parallel GAs on GPU

Many of early works of parallel GAs on GPUs take a naive approach: let each GPU thread run a sequential GA for a given portion of task, the same way used in parallel GAs on CPU to do a sequential GA over one processor. This approach, by ignoring the architectural feature of GPUs, cannot fully exploit the computing power of GPUs [6].

Because more than tens of thousands of threads run on GPUs, we cannot simply follow the old path of parallel GAs on CPU. Several issues need to be carefully examined according to the architectural feature of GPUs, such as how to arrange data access to maximize GPU memory bandwidth and how to organize threads to expose sufficient parallelism to GPUs in order to harness the power of GPUs [7].

This section first introduces briefly the basic concept of CUDA platform and then focuses on the issues of data layout, execution configuration, and memory access, which are the key enablers for parallel computing on GPU.

3.1. The CUDA platform

CUDA is a parallel computing platform and programming model developed by NVIDIA for general-purpose parallel computing on GPUs [8]. With CUDA, compute-intensive applications can be dramatically speeded up by harnessing the power of GPUs. CUDA is also a scalable programming model that enables programs to transparently scale their parallelism to GPUs.

In CPU-GPU heterogeneous system, conceptually, the sequential part of the workload runs on the CPU, while the compute-intensive portion of the application runs on thousands of GPU threads in parallel, as shown in **Figure 4**.

CUDA C is an extension of standard ANSI C with a handful of language extensions to enable parallel programming with a shallow learning curve for ones familiar with the C programming language. The CUDA Toolkit, provided in CUDA platform, includes GPU-accelerated libraries, a compiler, development tools, and the CUDA runtime, everything you need to develop GPU-accelerated applications.

The distinguishing features of CUDA C programming, compared with C programming on CPU, are that the CUDA platform exposes the memory model and execution model of GPUs to programmers to enable programmers to have more control over massive threads in order to optimize GPU implementation. The top three principles of CUDA C program optimization, listed in order of importance, are the following:

1. Exposing sufficient parallelism
2. Optimizing memory access
3. Optimizing instruction execution

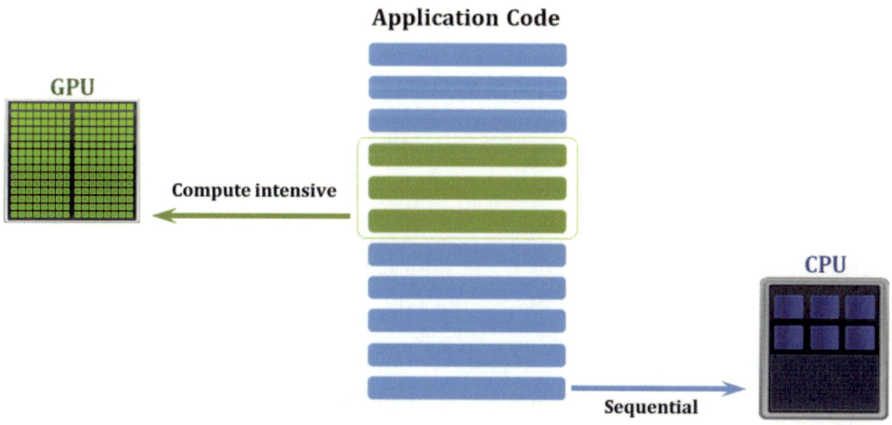

Figure 4. Acceleration on CPU-GPU heterogeneous system.

These principles are applicable almost to all kinds of computing architectures. The implementation of GPU kernels is architecture-dependent while applying these principles to harness the power of GPUs; therefore, it is necessary for a programmer to have some basic knowledge of the underlying architecture when parallelizing GAs on GPUs.

3.2. Granularity of parallel GAs on GPU

In parallel computing, the granularity of computations of an algorithm is a measure of the amount of computation which is conducted by that task [9]. Typically, the parallelism of a program can be classified into two categories: fine-grained and coarse-grained. In a fine-grained parallel algorithm, a program is broken down to a large number of small tasks, while in a coarse-grained parallel algorithm, a program is split into several large tasks. Both coarse-grained parallel GAs and fine-grained parallel GAs have been proposed for CPU architecture, but these parallel algorithms are too "coarse" to GPU architecture. Therefore, a new definition of granularity is necessary to describe parallelism for programs on GPU architectures [10].

The granularity for parallel GAs on GPU can be defined by how many GPU threads are working together to handle one chromosome during genetic operations. Basically, the parallelism can be classified into two categories:

1. The granularity in the chromosome level
2. The granularity in the gene level

If a solution in GAs' population is handled by one GPU thread independently, the granularity of the parallel GAs is in the chromosome level. If a solution is handled by a group of GPU threads cooperatively, the granularity of the parallel GAs is in the gene level. Another type of granularity of parallel GAs is a mixed form: some genetic operators in the chromosome level and some in the gene level. The granularity in the gene level also can be divided into several types: cooperative threads in block size or in warp size. In the former, a block of threads handles one solution, while in the latter, a warp of threads handles one solution.

GPUs take a partitioning approach to allocate compute resources among blocks and threads. The more a kernel requires the resources, the less the parallelism is exposed to GPUs. Therefore, the compute resource used in a kernel is a vital factor for the performance of a GPU kernel function, since more resource requirement will limit a kernel to expose sufficient parallelism to GPU.

Generally speaking, if a kernel of genetic operator is designed in the granularity of chromosome level, it will require more shared memory and local registers than a kernel designed in the granularity of gene level. For a large-scale application, it will become a major issue that hinders the performance. GPUs execute groups of threads in a unit known as warps in *single instruction, multiple thread* (SIMT) fashion; the kernel implemented in a granularity in the gene level will have more efficient access to global memory.

In most of the early works of parallel GAs on GPUs, the granularity of kernels belongs to the category of the chromosome level, because it is a nature extension of CPU program to GPU.

http://dx.doi.org/10.5772/intechopen.89152

It is also an easy way to migrate programs to GPU for the beginners of CUDA C programmers without knowing much in-depth knowledge of GPU architecture. This naive approach cannot fully utilize the computing power of GPU to accelerate GAs.

3.3. Data layout on global memory

Memory bandwidth and latency are key considerations in GPU applications. Bandwidth refers to the amount of data that can be moved to or from a given destination. Latency refers to the time the operation takes to complete. In designing a GPU kernel, we are primarily concerned about the global memory bandwidth. Most GPU kernels' performance is likely limited by memory accesses. To maximize global memory bandwidth, a vital step in kernel design is to make the kernel access global memory aligned and coalesced. How data are organized in the global memory will affect the way kernels access memory.

The population of GAs can be viewed logically as a 2D matrix with two dimensions: the index of chromosomes in the whole population and the index of gene within each chromosome. Physically GPUs store all data linearly in global memory. Therefore, there are two basic ways to layout the population in global memory:

1. The chromosome-based layout
2. The genotype-based layout

Let pop_size denote the population size and $nvar$ the total number of genes in one chromosome, the chromosome-based layout is shown in **Figure 5**, and the genotype-based layout is shown in **Figure 6**.

In the chromosome-based layout, the fast dimension is the index of gene, that is, one chromosome is allocated in a contiguous memory space. A kernel designed with the granularity of gene level can take the benefit of the coalesced access of global memory. In the genotype-based layout, the fast dimension is the index of chromosome in the population, and all i^{th} genes of all chromosomes are allocated to a contiguous range. Therefore, the stride between two consecutive genes in one chromosome is the number of pop_size. A kernel designed with the granularity of chromosome level may take the benefit of coalesced memory access. But this layout will be not good for genetic operations, such as evaluation and migration operation.

The discussion above is for the real-coded or integer-coded GAs. Conceptually, the data layout for the binary GA belongs to the chromosome-based layout. There are two kinds of implementation for the binary GAs:

1. Byte-wise binary method
2. Bitwise binary method

In the byte-wise encoding, 8 bits is used to represent 0 or 1, while in the bitwise method, only 1 bit is used to represent 0 or 1. A nature way to implement the binary GA is to pack multiple

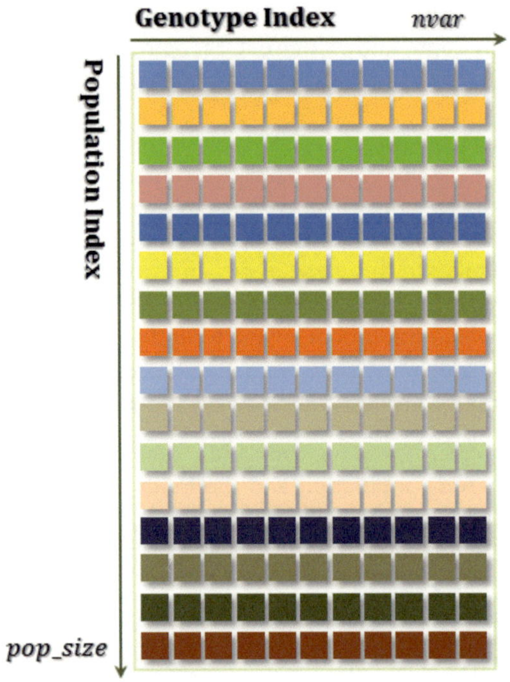

Figure 5. 2D view of the chromosome-based layout in global memory.

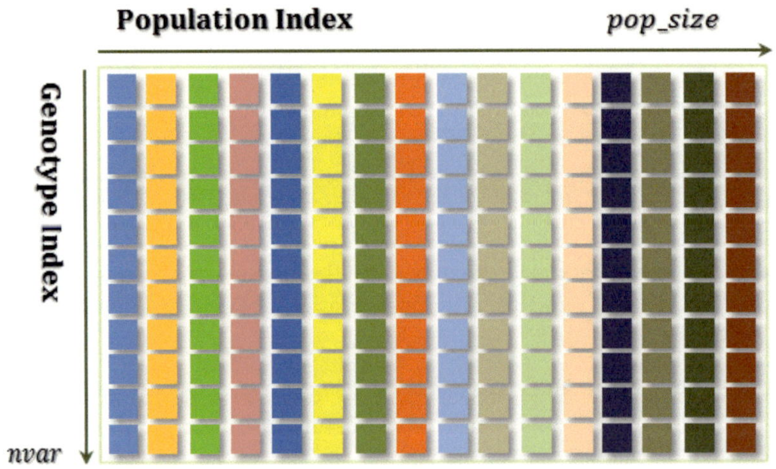

Figure 6. 2D view of the genotype-based layout in global memory.

bits into a non-Boolean data type, typically packing 32 bits into one unit for processing [11]. This bitwise approach can save global memory usage and also accelerate computation due to the less memory access.

3.4. Kernel execution configuration

The main difference between GPU and CPU programming is the level of programmer exposure to architectural features [7]. CUDA exposes the concepts of memory and thread hierarchy to grant the ability to control thread execution to a greater degree. GPU thread execution can be organized in a hierarchy structure: grid and block; therefore, programmers can configure the execution of a kernel function by specifying:

1. The number of threads per block and the dimension of the block
2. The number of thread blocks of a grid and the dimension of the grid

Grids and blocks represent a logical view of the thread hierarchy when executing a kernel function. Both block and grid can be arranged as 1D, 2D, and 3D layouts. How to configure a kernel execution heavily depends on the layout of data in the global memory in order to expose sufficient parallelism to GPU to saturate both instruction bandwidth and memory bandwidth. Exposing sufficient parallelism is the number one principle to optimize kernels, and grid and block heuristics play a vital role in kernel performance optimization.

3.5. Shared memory: the programmable cache

Shared memory can be viewed as a programmable cache and is a key enabler for kernel performance. By utilizing the shared memory, threads within the same block can easily cooperate with each other, facilitating the reuse of on-chip data and reducing greatly global memory traffic by caching data on-chip manually.

A programmer has the complete control over when data is moved into the shared memory and when data is evicted. CUDA makes it easier for you to optimize your application code by providing more fine control over data placement and on-chip data movement.

Sometimes, the non-coalesced accesses may not be easily avoided due to the nature of an algorithm. Shared memory provides a means to avoid non-coalesced memory accesses by staging data movement through the shared memory.

4. Parallel GAs on GPU

In this section, a concise review on selective works of parallel GAs on GPU is presented from the perspective of GPU computing. Several examples of combinatorial optimization problems are used to demonstrate on how to implement the master-slave model, island model, and cellular model on GPUs.

Although most published works are conducted on an early generation of GPU architecture, the basic idea on how to design parallel GAs on GPU is on the right track, and efficient implementations for different models share the common principles to optimize GPU performance, which is the focus of the chapter.

4.1. The master-slave model on GPU

The model of master-slave describes the relationship among multiple concurrent processes: one master process and many slave processes. The master process has the control over all other slave processes, distributing tasks among them and synchronizing them to work together. Each slave process does its own assigned task.

The earliest attempt at parallelizing GA on CPU is to let all slave processes calculate fitness values, the most time-consuming part of GA computation, while the master processes all other tasks. Lately, more tasks are moved into slave processes to accelerate the computation, as shown in **Figure 7**.

Because the relationship between GPU and CPU is the typical pattern of slave to master in nature, it is straightforward to implement the master-slave model on a GPU-CPU heterogeneous system. Earlier works take a very simple way to implement the model: GPU calculates the evaluation of the population only, while CPU does all other jobs, because evaluation takes major time comparing with other genetic operators. It is not an efficient implementation because it needs frequent data transfer between CPU and GPU at each generation of GAs, and the data movement between CPU and GPU dominates the execution time. Because most workloads are on CPU, it still is a bottleneck of performance, while GPU stays idle most time. Later works try to move more genetic operators onto GPU to speed up parallel implementations on GPU.

4.1.1. *Binary GA for one max problem*

One-Max problem is a toy problem in GA community just as the "Hello World" for C programmer. The problem is to produce a string of all 1 s, starting from a set of random binary strings. It has been widely used to test GA programs since it can well illustrate the potential of genetic algorithms.

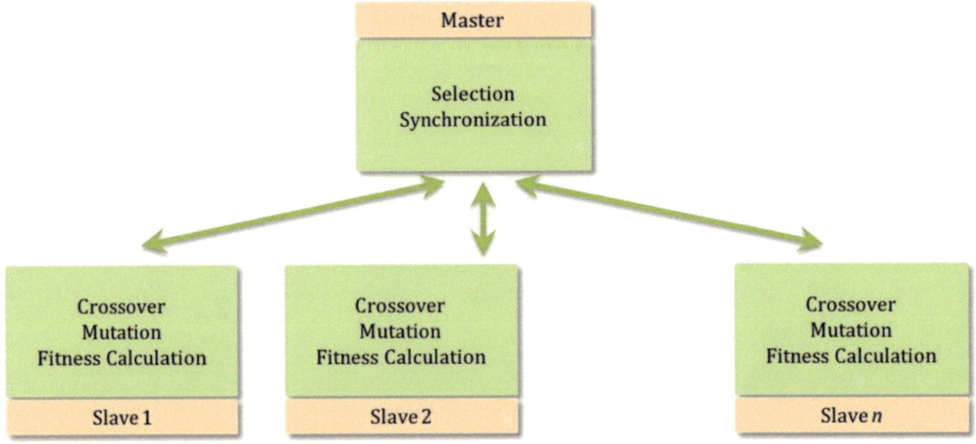

Figure 7. The master-slave model.

Debattisti et al. implemented a master-slave model of a fixed-length binary GA on GPU to solve the problem [12]. A byte-wise method is used to encode a solution: one byte is used to represent one bit of the genome, and a set of the unsigned char type data is used to encode a solution, that is, a string of binary value of 0 or 1. As mentioned in the last section, the bitwise encoding method is a better way to implement a binary GA on GPUs. Bitwise operations are available since CUDA 10.1 [13].

Except for initialization of the population, all other genetic operations, such as selection, crossover, mutation, and evaluation, are implemented on GPU. The crossover kernel implements the two-cut point method. Each thread block is responsible for dealing with two individuals to produce offspring, and each thread is responsible for handling 4 bytes each time. The mutation kernel implements the bitwise exclusive OR operations. Each thread block is responsible for dealing with one individual, and each thread is responsible for handling 4 bytes. All kernels are launched with 1D grid and 1D block configuration. Since multiple threads cooperatively deal with one individual, the granularity belongs to the gene level.

4.1.2. Real-coded GA for unconstrained optimization problems

Unconstrained optimization problems seek to maximize or minimize an objective function that depends on real variables without restrictions on these variables. For complex cases with many local optima, it highly depends on the initial point for conventional procedures to find the global optimum. Since GA is a population-based global search method, GAs can easily escape from local optimum during search process. That is why the unconstrained optimization problems are well tested in the GA community.

There are two typical ways to encode population of GAs: the binary-coded method and the real-coded method. For optimization problems, the real-coded approach outperforms the binary-coded approach [14].

Arora et al. implemented a master-slave model of the real-coded GAs on GPU to solve the unconstrained optimization problems [15]. The genotype-based layout is used to store the population in global memory, as shown in **Figure 6**.

The crossover kernel implements the simulated binary crossover (SBX) method, executing in the configuration of 1D block and 2D grid. Conceptually, each GPU thread works on a variable (or a gene) of two mates. Because thread blocks work independently with each other and one chromosome is operated by several different blocks, it will cause an issue that one chromosome may be selected to reproduce offspring many times.

The mutation kernel implements the one-variable change method using the execution configure of 1D block and 2D grid. Since one chromosome may be operated by multiple blocks independently, it has the same issue as what the crossover kernel has: one chromosome may be selected to reproduce offspring many times. The granularity of the implementation belongs to the gene level since multiple threads cooperatively deal with one or two individuals. Ellipsoidal function and Rosenbrock function are tested, which are well-used functions in GA literature.

4.2. Island model on GPU

In the island model, the population is divided into several subpopulations. By introducing geographical distribution of the whole population, the genetic diversity is hopefully preserved during the evolutionary process of different subsets. Each subset of population is an isolated breeding environment to evolve species locally. A new operator is introduced: the migration operator exchanges small portion of individuals among subsets to bring new genetic traits into each subset.

The island model is characterized by several factors: the number of islands, the topology of the migration, the interval of the migration, and the policy for selection and replacement during migration. The common migration topologies are given in **Figure 8**.

4.2.1. Binary GA for 0/1 knapsack problem

The 0-1 knapsack problem is a famous problem in combinatorial optimization: given sets of items, each with a weight and a value. The problem is which items should be put into a knapsack in order to maximize the total value without exceeding the capacity of the bag. Knapsack problems have been studied intensively, attracting both theorists and practitioners. The theoretical interest arises mainly from its simple structure, which allows exploitation of a number of combinatorial properties and permits more complex optimization problems to be solved through a series of knapsack-type subproblems. From a practical point of view, many industrial situations can be modeled as a knapsack problem. Since it is a well-known NP-hard problem, it has been well studied in GA community to test different implementations of GAs.

Jaros described implemented an island model of binary GA on multi-GPUs to solve the 0-1 knapsack problem [17]. The binary-coded method is used, since it is a nature representation for 0-1 binary knapsack problem. Each of the 32 items is packed into an integer data type; therefore, it is a kind of bitwise representation. The population is organized as a structure of arrays: the first array contains all chromosomes, and the second array contains the fitness values associated with each chromosome. The whole population is evenly distributed among multiple GPUs, and one GPU simulates one island to evolve a local population by using a steady-state method with elitism, the uniform crossover, the bit flip mutation, the tournament selection, and replacement. The unidirectional ring (1-way ring) is adopted as the migration topology where only adjacent nodes can exchange individuals. Migration among islands occurs after certain generations by exchanging the best individual and a number of randomly selected individuals. The tournament selection is used to pick emigrants. All the data exchanges among GPUs were implemented by means of MPI [18].

The chromosome-based layout is adopted to store the data in the global memory. The execution of all kernels is configured as 2D block and 2D grid. In the block layout, the fast dimension corresponds to the genes within a chromosome, while the slow dimension corresponds to different chromosomes, as shown in **Figure 9**.

The size of the fast dimension of 2D grid is fixed to 1, and the size of the slow dimension is set to the size of offspring divided by twice the slow dimension of 2D block, because two offspring were produced at once. Since each individual is processed by a warp, therefore, it belongs to the category of the granularity in the gene level.

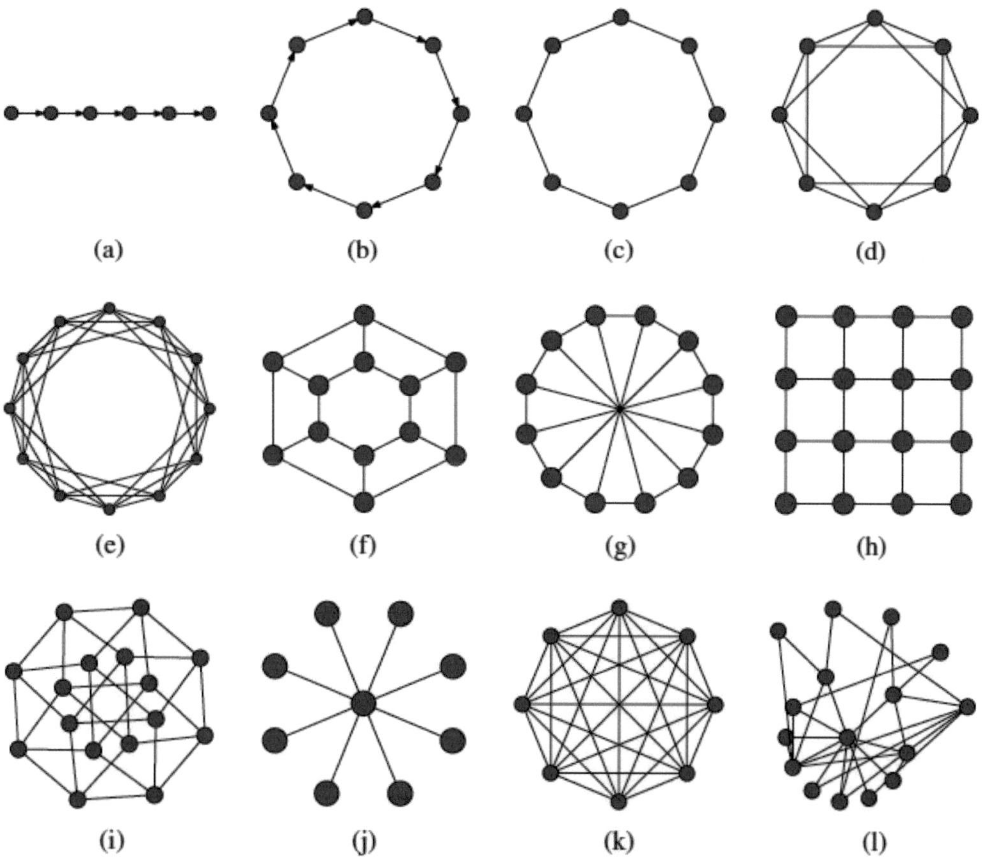

Figure 8. Common migration topologies (from [16]). (a) Chain, (b) I-way ring, (c) Ring, (d) +1+2 ring, (e) +1+2+3 ring, (f) Tours, (g) Cartwheel, (h) Lattice, (i) Hypercube, (j) Broadcast, (k) Fully-connected, and (l) Barabasi-Albert (3.2).

A knapsack instance with 10,000 items is used as a test case to simulate the real-world problem with a reasonable large data set. The tests are conducted on a two-node cluster, each equipped with seven GPUs, connected with InfiniBand.

4.2.2. Real-coded GA for unconstrained optimization problems

Luong et al. implemented an island model of real-coded GA for the unconstrained optimization problems [19]. The basic idea of their implementation is that each thread block is treated as an island, and each chromosome is handled by one thread for all operations of selection, crossover, mutation, and evaluation. Therefore, its granularity belongs to the chromosome level.

How data is organized in global memory was not described in the paper. If data is organized in the genotype-based layout, all kernels may take the merits of the coalesced memory access; if data is organized in the chromosome-based layout, all kernels will violate the principle of the coalesced memory access.

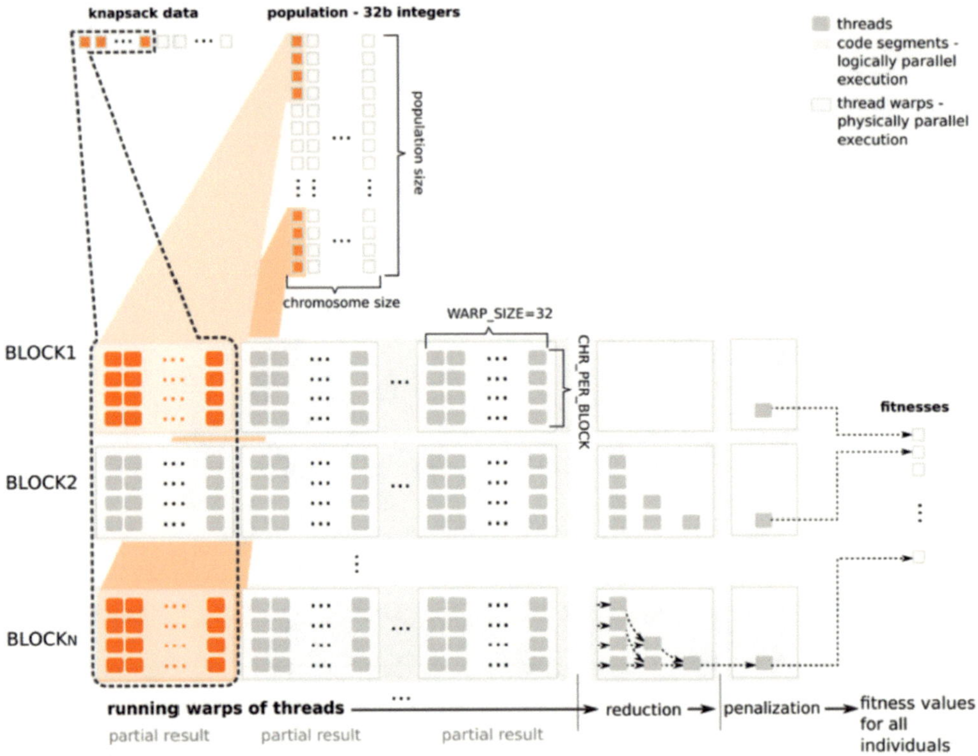

Figure 9. The configuration of kernel execution (from [17]).

The shared memory is used to staging data during the computation: all local subpopulations of each island stay in the shared memory. One major limitation of this way is the size of the memory, which makes only the cases with a very small population possible.

4.3. Cellular model on GPU

The cellular model imposes a geographic restriction on genetic search: individuals can only mate with its neighbors. The whole population is constricted with a spatial structure, defined as a connected graph. The common topology of the structure is a 2D toroidal mesh as shown in **Figure 10**, which limits the interactions between individuals. The use of cellular structure in populations helps for a better exploration over the search space with respect to the equivalent one with panmictic and decentralized populations.

The neighborhood of a given point in the mesh is defined in terms of Manhattan distance from it to its neighbor points. The two commonly used neighborhoods are L5, also called Von Neumann, and C9, also known as Moore neighborhood (Here, L stands for linear, while C stands for compact), as shown in **Figure 10**.

The neighborhood of all points is defined with the identical shapes. The neighborhood of each point of the mesh overlaps the neighborhoods of nearby points. The overlapping provides an

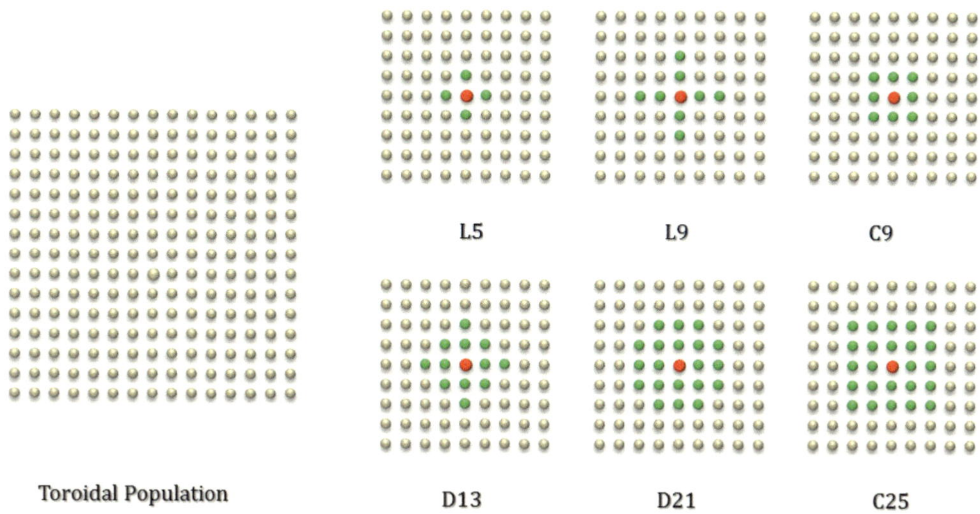

Figure 10. The toroidal mesh and neighborhoods.

implicit mechanism of migration to the cellular model: the genetic traits can spread smoothly through the overlapping neighborhoods. The mechanism helps the preservation of genetic diversity in the population longer than in a non-structured population and, therefore, provides a good trade-off between exploration and exploitation during the evolutionary process. The trade-off could be tuned by modifying the size of the neighborhood.

4.3.1. Integer-coded GA for quadratic assignment problem

The quadratic assignment problem is one of the fundamental combinatorial optimization problems in operations research. Conceptually, the problem is trying to find the optimal assignment of n facilities to n locations, knowing the distances between facilities and the flow between locations. Many combinatorial optimization problems can be written in this form. It has been a subject of extensive research in combinatorial optimization because of its importance in theoretical and practical domain. The problem is NP-hard, so there is no known algorithm for solving this problem in the polynomial time. Therefore, it becomes a target for GA community to show its stochastic and global search power for the complex optimization problems.

Cárdenas et al. implemented a cellular model of the integer-coded GAs for solving this problem [20]. Four types of neighborhoods, L5, C9, D17, and C25 (where L is linear, D is diamond, and C is compact), are mentioned, as shown in **Figure 11**.

All kernels are executed in a way that each GPU block corresponds to a chromosome and each GPU thread corresponds to a gene of the chromosome. Therefore, its granularity belongs to the gene level.

The crossover kernel implements the modified order crossover (MOX) method [21]. The first parent to MOX operator is decided by the block index, and the second parent is selected according to C9 neighborhood. The mutation kernel implements the random exchange method: selects two genes (two GPU threads) and exchanges their values.

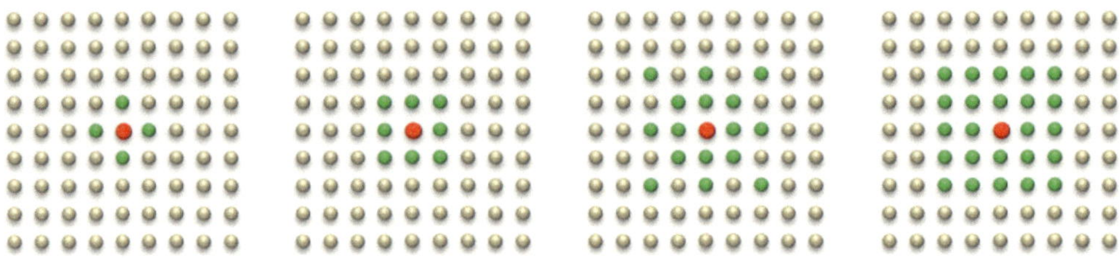

Figure 11. Neighborhood topologies.

Additional two nonstandard genetic operators are proposed: the transposition kernel and 2-opt kernel. In the transposition kernel, a portion of genes between two points is randomly generated in a chromosome and then reversed to obtain a new offspring. The 2-opt kernel mimics the 2-opt heuristic, a simple local search method to produce one offspring, both operators much like the mutation operator. It is a common practice to incorporate a kind of local search into genetic search to empower GAs when solving combinatorial optimization problems.

4.3.2. Integer-coded GA for independent task scheduling problem

Independent task scheduling problem is a kind of machine scheduling problem: assigning a set of independent computational tasks onto the different processors in a heterogeneous cluster. Finding a schedule that minimizes makespan to the problem is known to be NP-complete. The background of the problem is how to assign independent tasks onto the different processors in a cluster; therefore, time constraint is vital to this problem; usually, a limited amount of time is available for calculating the task schedule.

Pinel et al. implemented a cellular model of the integer-coded GA to solve the problem [22]. The population is arranged into a 2D toroidal mesh. The solution to GAs is encoded as a string of integer: the index of the string corresponds to a task, and the integer for a given index in the string corresponds to the machine to which this task is assigned. Uniform proportional recombination (UPR) is implemented as the crossover kernel, as shown in **Figure 12**.

The circled task in the middle string is the task being updated. Its neighborhood is defined by L5, the tasks in above, below, right, and left strings. The offspring is generated by the winner among the possible solutions according to its neighborhood. Two different criteria are defined for choosing the winner. The crossover operator is implemented with two kernels: one kernel is used to compute the probability for each solution in the neighborhood, and the other kernel is used to update tasks to generate offspring.

The crossover operator runs in a way that one thread handles with one task of a solution. Therefore, its granularity belongs to gene level. The other genetic operators of mutation, fitness, and replacement kernels, are launched in a way of one thread per solution, which belongs to the chromosome level.

Several instances are tested ranging from 512 tasks over 16 machines to 65,536 tasks over 2048 machines.

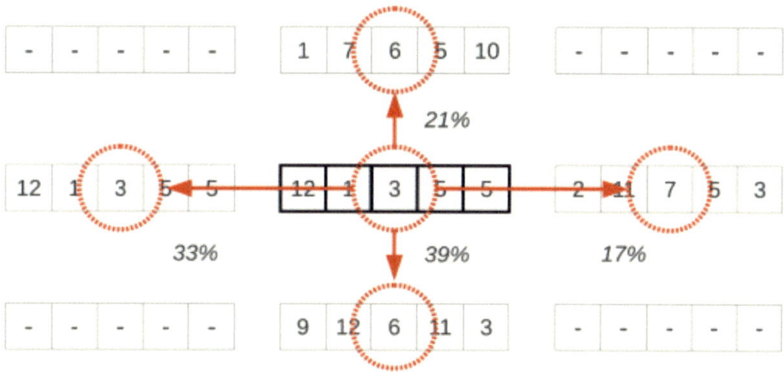

Figure 12. Illustration of crossover operator (from [22]).

5. A case study: multi-objective GAs for railway scheduling

The railway scheduling problem consists of building the timetable of trains, moving on a railway network under certain constraints. Nitisiri et al. reported their works of parallel and multi-objective GA implementation on GPU to generate train schedule for the Bangkok mass transit system (BTS) [23].

Bangkok BTS is a double-track railway system, and each track is operated in a single direction, as shown in **Figure 13**.

The BTS Silom Line consists of 13 stations, starting at National Stadium (W1) in central Bangkok and traveling southward to the last station S12 at Bang Wa station with the total length of 13 kilometers. The average number of passengers during the peak time in 2015 is above 2000 peoples. The BTS system operates daily from 05:00 am to 24:00 pm. The regular timetable is shown in **Figure 14**; the headways during normal operation are 15 mins. The total number of operating cycle is 91 cycles.

The train scheduling problem can be mathematically formulated as multi-objective integer programming problem with complex constraints. Two objectives are involved in this scheduling: (1) minimize average waiting time for the passenger, and (2) minimize a total number of the operating cycles during the operating period. These objectives are in conflict with each other: the more trains operating during the day, the less waiting time for the passenger but increasing the operational cost for the company.

A schedule is encoded as a $C \times 2S$-length matrix G, where C is the number of train operating cycles and S is the number of the stations, as shown in **Figure 15**.

A master-slave model is implemented on GPU for multi-objective GA with hybrid sampling strategy and learning-based mutation to solve the railway train scheduling problem. Data layout in the global memory belongs to the chromosome-based layout. Two-point cut crossover and learning-based mutation are used to perform genetic search. For evaluating each individual with respect to two objectives, a method called Pareto dominating and dominated

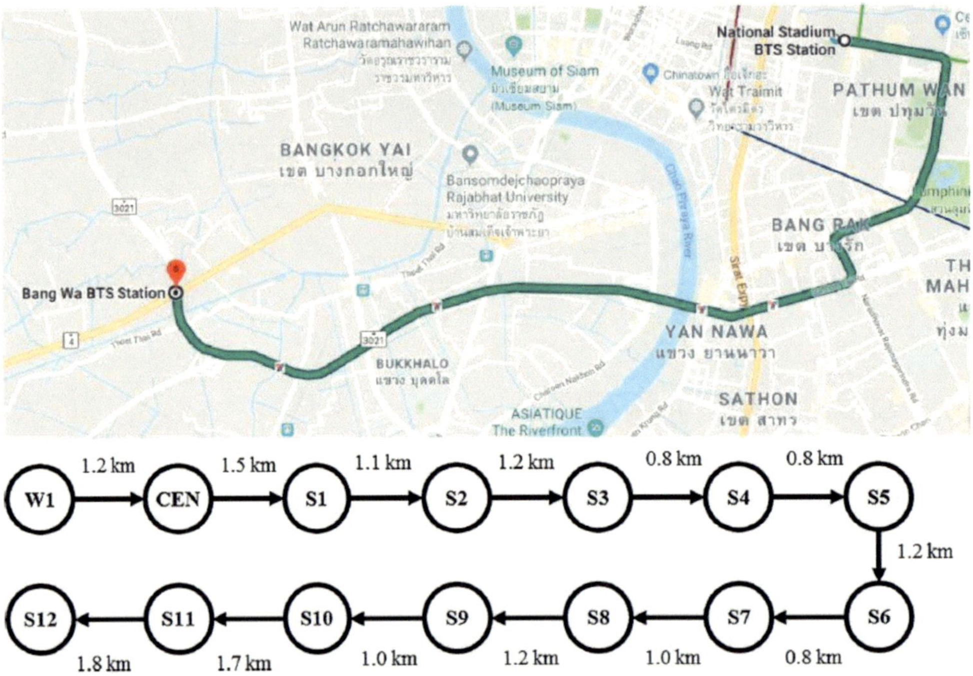

Figure 13. The Bangkok BTS transit line.

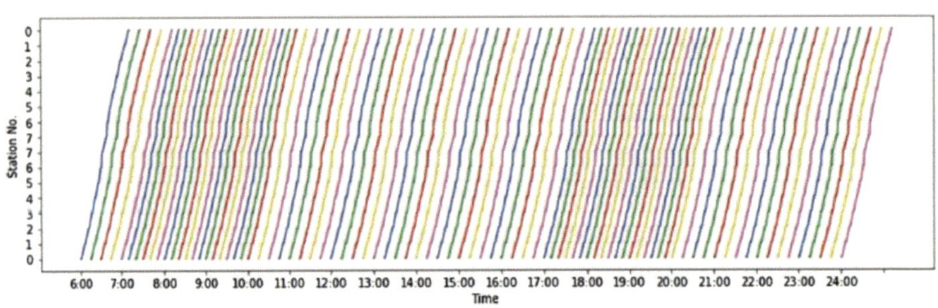

Figure 14. Regular timetabling scheme for Bangkok BTS transit line.

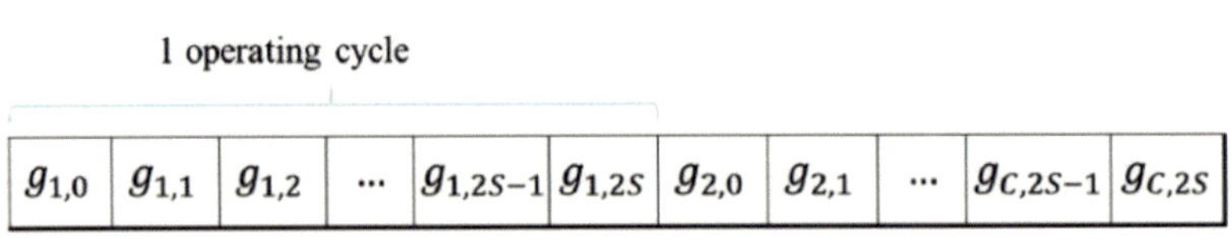

Figure 15. Genetic representation of a schedule.

relationship-based fitness function (PDDR-FF) is implemented as the evaluation kernel, which prefers the solutions in the edge area of the Pareto front [24]. A hybrid sampling strategy is adopted to select next generation, which has a tendency to converge toward the central area of the Pareto front [25].

Kernels are launched with 2D grid and 3D blocks and one GPU thread healing with one chromosome; therefore, its granularity belongs to the type of the chromosome level.

6. Discussion

Most published works on parallel GAs on GPU are conducted before 2013 on the earlier generations of GPU architecture. Since 2008, six generations of Tesla architectures for high-performance computation are released: Tesla, Fermi, Kepler, Maxwell, Pascal, and Volta. The average life span of one generation is about 2 years. The latest versions of GPU are Tesla V100 and Tesla P100, delivering a unified platform for accelerating both HPC and AI works.

Most published works tackle relatively small-scale problems. In handling large-scale problems, an efficient implementation of parallel GAs on multiple GPUs becomes imperative. A promising approach to do this is the hybrid model of island and cellular model on modern GPUs.

The earliest effort to combine these two models is given by Gruau in 1194, implemented in CPU to evolve and train a neural network [26]. Dorronsoro et al. reported their research works to use the hybrid model to solve a very large scale of vehicle routing problem in 2007, executed on a grid composed of 125 heterogeneous CPU nodes [27].

It is worthy to make further investigation on this hybrid model over multiple GPUs: implement a cellular mode in each GPU and treat each GPU as an island, as shown in **Figure 16**, where a 2D torus migration topology is defined among GPUs.

The hybrid model possesses both merits of island and cellular model: keeping the genetic diversity of population among islands and empowering local search ability on a structured population within each island. Implementing parallel GAs on multiple GPUs enables us to tackle larger-scale problems.

One of the promising areas to apply this hybrid model is to evolve better deep neural network. The notorious issue with deep net is what is called hyper-parameters, the very values of the deep net that cannot be learned and must be predetermined subjectively. Ever since the late 1990s, researchers have tried to evolve the best structure for network with GAs to solve the hyper-parameter issue [28]. There is a special term for this research that combines two powerful AI algorithms of genetic algorithms and neural networks: neuroevolution. Since 2016, the idea of architecture search through neuroevolution is attracting a number of researchers. Gaier and Ha reported their interesting results on using evolutionary algorithms to find minimal neural network architectures that can perform several reinforcement learning tasks without weight training [29]. For a large-scale application, weight training will be a painful and time-consuming process for most deep learning network methods.

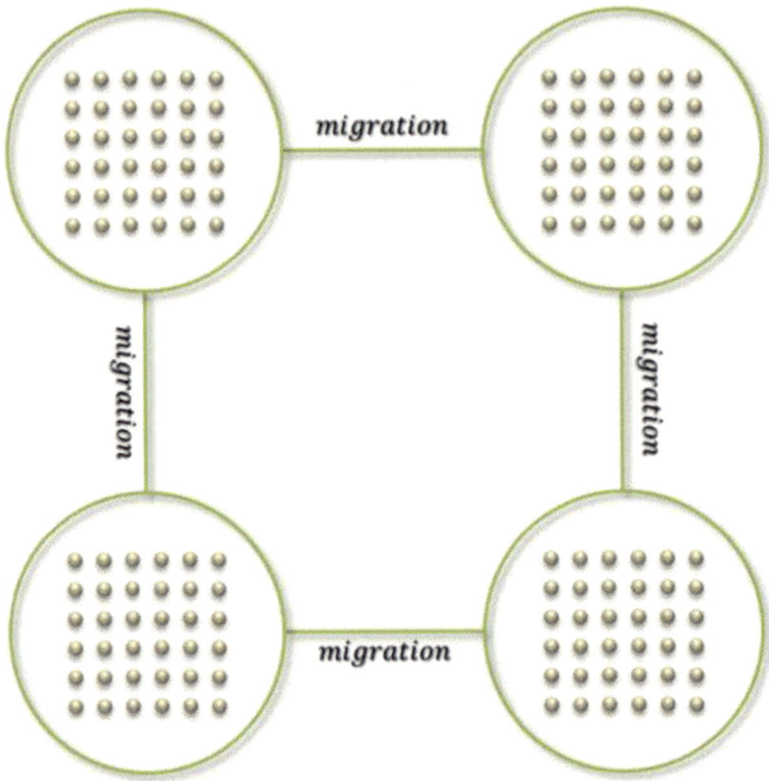

Figure 16. An example of a hybrid model over multiple GPUs: each GPU is an island, a cellular model is implemented within each GPU, and a 2D torus migration topology is defined among GPUs.

7. Conclusion

This chapter provides a concise yet comprehensive overview of selective works of parallel GAs on GPU from the perspective of GPU computing architecture.

Nowadays, the CPU-GPU heterogeneous architecture has become the mainstream computing platform that accelerates applications, which has led to a fundamental paradigm shift in parallel programming. The difference between the parallel GAs on CPU and the parallel GAs on GPU is clearly distinguished in the chapter. Conceptually, a parallel GA on CPU runs multiple instances of a sequential GA over several processors, while a parallel GA on GPU organizes massive threads of hundreds of thousands to work truly in parallel to exert the power of GPU computing.

Early tries to accelerate GAs with GPU computing takes a naive approach: let each GPU thread run a sequential GA, which is not the way we should take to design a parallel program over such massive threads of GPU computing architecture. In the chapter, the parallel model, the design approaches, and implementation strategies of parallel GAs are examined from the perspective of GPU computing architecture. Although designing a parallel GAs on GPU is both problem-dependent and architecture-dependent, the efficient implementations of parallel GAs on GPUs share the common principles of kernel optimization: (1) exposing sufficient parallelism, (2) optimizing memory access, and (3) optimizing instruction execution. CUDA platform exposes GPU architectural features directly to programmers that enables us having more control over the massively GPU threads in order to fully exploit the computing power of GPUs.

Most research works on parallel GAs on GPU are published around 2013 on the earlier generations of GPU architecture. Since then, several generations of Tesla architectures for high-performance computation are released. The latest versions are Tesla V100 and P100. It would be a very promising research area to probe all these parallel models on the latest generations of GPU architectures to find out how can we accelerate parallel GAs on them.

The research on hybridizing island and cellular model on GPU clusters would be a very promising area that enables us to tackle hyper-scale computing applications. The hybrid model have both merits of island and cellular models: keeping the diversity of solutions among islands and empowering the ability of local search onto the global and stochastic search within each island. It is worthy to try this model for neuroevolution in deep learning applications. Now neuroevolution is making a comeback, and GPU-accelerated parallel GAs will make this happen certainly.

Author details

John Runwei Cheng[1]* and Mitsuo Gen[2]

*Address all correspondence to: runweicheng@gmail.com

1 Broad Geophysical Technology, Inc., Houston, TX, United States

2 Research Institute of Science and Technology, Tokyo University of Science and Fuzzy Logic Systems Institute, Japan

References

[1] Gen M, Cheng R. Genetic Algorithms and Engineering Design. New York: John Wiley & Sons; 1997. p. 427

[2] Gen M, Cheng R. Genetic Algorithms and Engineering Optimization. New York: John Wiley & Sons; 2000. p. 511

[3] Goldberg DE. Genetic Algorithms in Search, Optimization and Machine Learning. Reading: Addison-Wesley; 1989. p. 432

[4] Gen M, Cheng R, Lin L. Network Models and Optimization: Multi-Objective Genetic Algorithms Approach. New York: Springer; 2008. p. 692

[5] Luque G, Alba E. Parallel Genetic Algorithms: Theory and Real-World Applications. Berlin: Springer; 2011. p. 95

[6] Zheng L, Lu Y, Ding M, Shen Y, Guo M, Guo S. Architecture-based performance evaluation of genetic algorithms on multi/many-core systems. In: Proceedings of 14th IEEE International Conference on Computational Science and Engineering (CSE 2011). 24-26 August 2011. Dalian, China: IEEE Computer Society; 2011. pp. 321-334

[7] Cheng J, Grossman M, McKercher T. Professional CUDA C Programming. New Jersey: John Wiley & Sons; 2014. p. 497

[8] CUDA Zone. Available from: http://developer.nvidia.com/cuda-zone [Accessed on: 18-06-2019]

[9] Hwang K. Advanced Computer Architecture: Parallelism, Scalability, Programmability. New York: McGraw-Hill Higher Education; 1992. p. 771

[10] Cheng JR, Gen M. Accelerating genetic algorithms with GPU computing: A selective overview. Computers and Industrial Engineering. 2019;**128**:514-525. DOI: 10.1016/j.cie.2018.12.067

[11] Pedemonte M, Alba E, Luna F. Bitwise operations for GPU implementation of genetic algorithms. In: Proceedings of Genetic and Evolutionary Computation Conference (GECCO-2011). Ireland: Dublin; 2011. pp. 439-446

[12] Debattisti S, Marlat N, Mussi L, Cagnoni S. Implementation of a simple genetic algorithm within the CUDA architecture. In: Proceedings of Genetic and Evolutionary Computation Conference (GECCO-2009). Montreal, Canada; 2009

[13] CUDA Toolkit Documentation v10.1.168. Available from: https://docs.nvidia.com/cuda/index.html [Accessed on: 20-06-2019]

[14] Michalewicz Z. Genetic Algorithm + Data Structure=Evolution Programs. 3rd rev ed. New York: Springer; 1996. p. 387

[15] Arora R, Tulshyan R, Deb K. Parallelization of binary and real-coded genetic algorithms on GPU using CUDA. In: Proceedings of IEEE Congress on Evolutionary Computation (CEC 2010). Spain: Barcelona; 2010. pp. 1-8

[16] Izzo D, Ruciński M, Biscani F. The generalized Island model. In: de Vega FF et al., editors. Parallel Architectures & Bioinspired Algorithms. Berlin: Springer-Verlag; 2012. pp. 151-169

[17] Jaros J. Multi-GPU Island-based genetic algorithm for solving the knapsack problem. In: IEEE Congress on Computational Computation (CEC 2012). Brisbane, Australia; 2012. pp. 1-8

[18] OpenMPI. Open source high performance computing. Available from: http://www.open-mpi.org/ [Accessed on: 18-06-2019]

[19] Luong TV, Melab N, Talbi E. GPU-Based Island model for evolutionary algorithms. In: Proceedings of the Genetic and Evolutionary Computation Conference (GECCO-2010). Portland, Oregon; 2010. pp. 7-11

[20] Cárdenas GE, Poveda Ch R, García HO. A solution for the quadratic assignment problem (QAP) through a parallel genetic algorithm based grid on GPU. Applied Mathematical Sciences. 2017;**11**(57):2843-2854

[21] Wroblewski J. Theoretical foundations of order-based genetic algorithms. Fundamenta Informaticae. 1996;**28**(3-4):423-430

[22] Pinel F, Dorronsor B, Bouvry P. Solving very large instances of the scheduling of independent tasks problem on the GPU. Journal of Parallel and Distributed Computing. 2013;**73**(1):101-110

[23] Nitisiri K, Gen M, Ohwada H. A parallel multi-objective genetic algorithm with learning based mutation for railway scheduling. Computers and Industrial Engineering. 2019;**130**:381-394. DOI: 10.1016/j.cie.2019.02.035

[24] Gen M, Cheng JR, Nitisiri K. Advances in hybrid genetic algorithms with learning and GPU for scheduling problems. In: The International Conference for High Performance Computing (SC19). Denver, Colorado; 2019. pp. 17-22

[25] Gen M, Lin L. Genetic algorithms. In: Wah B, editor. Wiley Encyclopedia of Computer Science and Engineering. Hoboken, N.J.: John Wiley & Sons; 2009. pp. 1367-1381

[26] Gruau F. Neural network synthesis using cellular encoding and the genetic algorithm [Unpublished doctoral dissertation]. L'Universite Claude Bernard-Lyon I. 1994

[27] Dorronsoro B, Arias D, Luna F, Nebro A, Alba E. A grid-based hybrid cellular genetic algorithm for very large scale instances of the CVRP. In: High Performance Computing & Simulation Conference (HPCS 2007). 2007. pp. 759-765

[28] Yao X. Evolving artificial neural networks. Proceedings of the IEEE. 1999;**87**(9):1423-1447

[29] Gaier A, Ha D. Weight agnostic neural networks. Available from: https://arxiv.org/abs/1906.04358

Chapter 7

Digital Twin Technology

Zongyan Wang

Additional information is available at the end of the chapter

http://dx.doi.org/10.5772/intechopen.80974

Abstract

Digital twin technology is considered to be the core technology of realizing Cyber-Physical System (CPS). It is the simulation technology that integrates multidisciplinary, multiphysical quantity, multiscale and multi probability by making full use of physical model, sensor update, operation history and other data. It is the mapping technology for the whole lifecycle process of physical equipment in virtual space. It is the basic technology of Industrial 4.0. This chapter mainly introduces: (1) the generation of digital twin technology; (2) the definition and characteristics of digital twin technology; (3) the relationship between digital twin and digital thread; (4) the implementation of the product digital twin model; and (5) the research progress and application of digital twin research.

Keywords: digital twin, digital twin models, digital twin workshops, digital twin applications, Cyber-Physical System (CPS), virtual reality fusion, intelligent manufacturing, products lifecycle management, modeling and simulation

1. The generation of digital twin technology

The use of the "twin/twins" concept in the manufacturing can be traced back to NASA's Apollo program [1]. In the project, NASA needs to make two identical spacecraft. The aircraft left on earth is called a twin and is used to reflect the status/condition of the space vehicle in action. During the flight preparation, the space vehicle known as the twins is widely used in training. During the mission, the twins were used to simulate the space model on the ground, and it can accurately reflect and predict the status of the space vehicle in operation as much as possible, so as to assist the astronauts in orbit to make the most correct decision in emergencies. From this perspective, it can be seen that the twins are actually a prototype or model that reflects the real operation situation in real time through simulation. It has two significant characteristics: (1) the twins with the objects to be reflected are almost exactly the same as the appearance (the geometry and size of the product), content (the structure of the product and its macro- and microphysical properties) and properties (the function and performance of the product); and (2) it allows you to mirror/reflect real operation/state by means of simulation, etc. It needs to be pointed out that the twin at this time is still physical.

In 2003, professor Michael Grieves proposed the concept of virtual digital representations equivalent to physical products in a product lifecycle management (PLM) course at the University of Michigan and gives it a definition: which is a digital copy of one or a set of specific devices that can abstractly represent a real device and can be used as a basis for testing under real or simulated conditions [2]. The concept stems from the expectation of a clearer expression of the information and data of the device, hoping to put all the information together for a higher level of analysis. Although this concept was not called at the time as digital twin model (from 2003 to 2005, known as the "mirrored spaced model" [3], and during 2006–2010, known as "information mirror model" [4]), but its conceptual model had all the elements of the digital twin model, namely the physical space and virtual space and the relation or interface between them, therefore, it was regarded as the embryonic form of digital twin.

In 2006, the US National Science Foundation (NSF) first proposed the concept of the information physics system Cyber-Physical Systems (CPS), and could also be translated into a network entity system, or an information physical integration system [5]. The information physics system is defined as a network composed of physical input and output and interactive components. It is neither different from the independent equipment that is not connected to the network, nor it is different from the pure network without physical input and output.

In 2011, professor Michael Grieves, in his book "Virtually perfect: driving innovative and lean products through product lifecycle management" [6], cited the conceptual model of the noun digital twin model (digital twin), which is described by his co-author John Vickers, and it is still in use today. Its conceptual model is shown in **Figure 1**, including three main parts: (1) real space entity products; (2) virtual space virtual products; and (3) data and information interface between real space and virtual space.

This conceptual model greatly expands the "twins" in the Apollo program: (1) it digitizes the twin model and use digital expression to build a virtual product with the same content and nature as the product entity in appearance; (2) it introduces virtual space and establishes the association between virtual space and real space, so that data and information can be

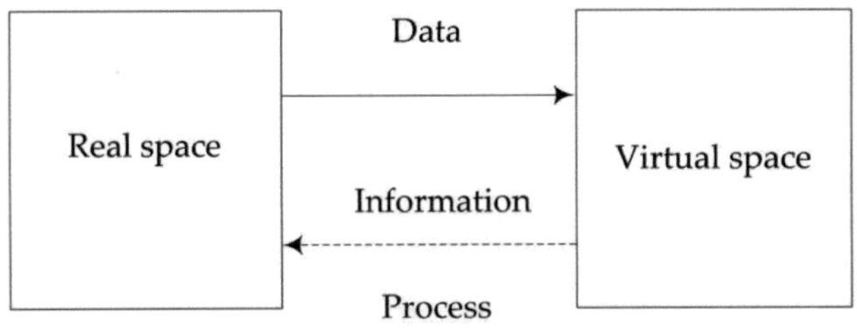

Figure 1. Conceptual model of digital twin.

exchanged between each other; (3) it visually reflects the concept of integrating the real with the imaginary, and controlling the real with the imaginary; and (4) extension of the concept and extension, in addition to products, factory, workshop, production lines, manufacturing resources (work position, equipment, personnel, materials, etc.), in the virtual space corresponding digital twin model can be set up.

However, it did not attract much attention from scholars when the concept and model were proposed in 2003. The main reasons are: (1) there were limited technical means to collect product-related information in the production process at that time, most of which were based on manual methods and paper documents; in particular, it is difficult to realize the on-line real-time collection of production data; (2) the digital description of physical products is not yet mature, and relevant software and hardware cannot support the precise definition and description of related properties and behaviors of physical products in virtual space; and (3) at that time, it was difficult to realize real-time processing of big data with computer performance and algorithm, and mobile communication technology was not mature enough, and real-time data transmission between virtual and real data was difficult to achieve.

After 2011, the digital twin ushered in a new development opportunity. The digital twin was proposed and further developed by the US Air Force Research Laboratory in 2011; the aim is to solve the maintenance and life prediction of aircraft in the future complex service environment [7]. In 2025, they plan to deliver a new type of space vehicle and digital model that corresponds to the physical product. The digital twin has super realism in two aspects: (1) it includes all geometric data, such as machining error and (2) includes all material data, such as material microstructure data. In 2012, the US Air Force Research Laboratory proposed the concept of Airframe Digital Twin" as the hyper-realistic model of the airframe being manufactured and maintained. The Airframe Digital Twin can be used to simulate and judge whether the airframe meets the task condition. It is an integrated model composed of many sub-models [8].

The Airframe Digital Twin is a consistent model and computational model of a single airframe in the whole product lifecycle. It is associated with the materials, manufacturing specifications and processes used to manufacture and maintain aircraft. It is also a sub-model of aircraft digital twin, which is an integrated model including electronic system model, flight control system model, propulsion system model and other subsystem models. At this time, digital twin enters the initial planning and implementation stage from the conceptual model stage, and its connotation and nature are further described and studied.

Specifically, (1) it highlights the hierarchical and integrated nature of digital twin, for example, the aircraft digital twin, the airframe digital twin, airframe structure model, material state evolution model and so on, and is beneficial to the gradual implementation and final realization of digital twin; (2) it highlights the hyperrealism of digital twin, including geometric model, physical model, material evolution model, etc.; (3) it highlights the universality of digital twin, that is, it includes the whole product life cycle and extends from the design stage to the subsequent product manufacturing stage and product service stage; (4) it highlights the consistency of digital twin in the whole life cycle of products, which reflects the idea of a single data source; and (5) it highlights the computability of digital twin, and the real state of corresponding product entity can be reflected in real time through simulation and analysis.

In 2012, in the face of future aircraft light quality, high load and the demand of the longer service time under more extreme environment, NASA and the US Air Force Research Laboratory in cooperation put forward the common digital twin example of future aircraft. For aircraft, flight systems or launch vehicles, they define digital twin as an integrated multi-physical, multi-scale, probabilistic simulation model for aircraft or system, which uses the best available physical models, updated sensor data, and historical data to reflect the state of the flying entity corresponding to the model [9]. In the same year, digital twin were formally introduced into the public view in the roadmap of modeling, simulation, information technology and processing released by NASA [10]. The definition can be considered as a periodic summary of previous research by the US Air Force Research Laboratory and NASA, especially it emphasizes the integration, multiphysical, multiscale, probabilistic characteristics of digital twin, and its main function is to be able to reflect the state of corresponding flight products in real time (continuing the function of the twins of the early Apollo project); the data used includes the current best available product physical models, updated sensor data, and historical data for product groups.

In 2013, Germany proposed the "Industry 4.0," whose core technology is Cyber-Physical System (CPS). CPS is a multidimensional complex system which is integrated with computing, communication, control, network and the physical environment [11]. Based on the big data network and mass computing and through the organic integration and deep cooperation of 3C (computing, communication, control) technology, the real-time perception, dynamic control and information service of large-scale engineering systems can be realized. CPS from physical space, environment, activities, large data collection, storage, modeling, analysis, mining, evaluation, prediction, optimization and coordination combine with design of the object, testing, and performance characterization to realize the depth fusion of network space (cyberspace) and physical space, real-time interaction and mutual coupling and update on each other; Furthermore, it promotes the comprehensive intelligence of industrial assets through self-perception, self-memory, self-cognition, self-decision-making, self-reconstruction and intelligent support. CPS connects people, machines and things. The virtual and real bidirectional dynamic connection in CPS connotation has two steps: (1) virtual entity such as the design of a product is simulated first and then manufactured and (2) entity virtualization. In the process of manufacturing, using and running, entities reflect their status to the virtual end and conduct monitoring, judgment, analysis, prediction and optimization through the virtual mode.

By constructing a closed loop channel for data interaction between information space and physical space, CPS can realize the interaction between information virtual model and physical entity. The emergence of digital twin provides a clear idea, method and implementation way for CPS. On the basis of the physical entity modeling the static model, through the real-time data acquisition, data integration and monitoring, dynamic tracking of physical entity working status and progress measurement result (such as acquisition, traceability information, etc.), the physical entities in the physical space are reconstructed in the information space, forming digital twin with the ability of perception analysis and decision execution. Therefore, from this perspective, digital twin is the core technology of CPS.

In 2014, Professor Michael Grieves elaborated digital twin in detail in his white paper "Digital Twin: Manufacturing Excellence through Virtual Factory Replication" [2]. In the same year, the U.S. defense department, PTC, Siemens and Assaults accepted the term "digital twin" and began to use it in marketing campaigns. It needs to be pointed out that they all use "digital twin" instead of "digital twins".

In 2015, General Electric Company planned to implement real-time monitoring, timely inspection and predictive maintenance of engines based on digital hygiene and through its own cloud service platform—Predix, using advanced technologies such as big data and Internet of Things [12]. In Made in China 2025, CPS is considered to be an integrated technical system that supports the deep integration of industrialization and informatization, and is an important starting point for promoting the integration of manufacturing and Internet.

In 2017, in order to realize the interactive fusion of the physical world and the information world of the manufacturing workshop, Tao Fei puts forward the realization mode of the digital twin workshop, and made clear its system composition, operation mechanism, characteristics and key technology, which provided the theory and method reference for the realization of the information physical system of the manufacturing workshop [13]. For two consecutive years (2016 and 2017), Gartner, the world's most authoritative IT research and consulting firm, listed digital twin as one of the top 10 strategic technology trends of the year. The world's largest weapons manufacturer Lockheed Martin in November 2017 ranked digital twin as the top of the six leading technology in the future defense and aerospace industries. In December 2017, the China Association for Science and Technology (CAST) Intelligent Manufacturing Academic Union (CIMA) listed digital twin as one of the top 10 technological advances in intelligent manufacturing at the World Intelligent Manufacturing Conference [14].

It can be seen that the digital twins have developed rapidly in both theoretical and application levels in recent years. At the same time, the application range has gradually shifted from the product design stage to the product manufacturing stage and operation and service stage, which has attracted wide attention of scholars and enterprises. The main reasons are the following aspects:

1. the rise and wide application of model digital expression technology such as model lightweight, MBD, physics-based modeling, etc., makes it possible to accurately describe physical products at various stages of the product life cycle using digital methods; and

2. the rapid popularization and application of the new generation of information and communication technologies such as large data, Internet of Things, mobile Internet, cloud computing, and the rapid development of computer science and technology such as large-scale computing, high-performance computing, distributed computing, as well as the emergence of intelligent optimization algorithms, such as machine learning and deep learning, make products reliable with real-time dynamic data collection and predict possible, such as fast transmission, storage, analysis and decision, and provide important technical support for real-time correlation and interaction between virtual space and physical space.

2. The definition and characteristics of digital twin technology

From the perspective of the origin and current development of digital twin, its applications mainly focus on product design and operation and maintenance stages, but with the rapid spread and application of new generation of information and communication technology such as big data, Internet of Things, mobile Internet, and cloud computing, digital twin have gone beyond the traditional product design and operation phases. To make it easier to understand the digital twin, this section gives a definition of the digital twin technology.

2.1. Definition of digital twin and digital twin model

Digital twin refers to the processes and methods for describing and modeling the characteristics, behavior, formation process, and performance of physical objects using digital technology, and can also be referred to as digital twin technology. The digital twin model refers to a virtual model that completely corresponds to and is consistent with the physical entities in the real world, and can simulate its behavior and performances in a real-time environment in real time. It can be said that digital twin is techniques, processes, and methods, and that digital twin models are objects, models, and data. Digital twin technology is not only using human theory and knowledge to build virtual models but also can use virtual model simulation technology to explore and predict the unknown world, to find better ways and means, and constantly inspire human innovative thinking. The pursuit of optimization and progress, therefore, in digital twin technology provides new ideas and tools for current manufacturing innovation and development.

In the future, there will be a digital twin model in the virtual space that is exactly the same as the entity in the physical space. For example, the physical factory has a corresponding factory digital twin model in the virtual space, and the physical workshop has a corresponding workshop digital twin model in the virtual space. Physical production lines in the virtual space have corresponding production line digital twin model and so on.

Digital twin is the foundation of the intelligent manufacturing system. The most important enlightening significance of digital twin is that it realizes the feedback from the physical system to the digital model of cyberspace [15].

This is a feat of reverse thinking in the industrial field. People try to plug everything that happens in the physical world back into digital space. Only full life tracking with loop feedback is the true full lifecycle concept. In this way, the digital and physical world can be truly harmonized throughout the entire lifecycle. Various types of simulation, analysis, data accumulation, and mining based on digital models, and even the application of artificial intelligence, can ensure its applicability to real-world physical systems. This is the significance of digital twin to intelligent manufacturing. The intelligence of an intelligent system must be first perceived, modeled, and then analyzed. Without digital twin' accurate modeling of a real production system, the so-called intelligent manufacturing system is passive water and cannot be implemented.

2.2. Definition of product digital twin model

Considering the evolution process and related explanations of the existing product digital twin model, the author gives the definition of the product digital twin model: the product digital twin model refers to the full-element reconstruction and digitized mapping of the physical entity's working state and work progress in the information space, and is an integrated multiphysics, multiscale, hyperrealistic, dynamic probability simulation model that can be used for simulating, monitoring, diagnosing, predicting, and controlling the formation process, state, and behavior of physical entities in the real world. Product digital twin model generated by the product model based on the product design stage, and during the next product manufacturing and service stage, with the product data and information interaction between physical entities, constantly improves their integrity and accuracy, finishing a complete and accurate description of product physical entity. Some scholars have also interpreted the digital twin model as digital mirror, digital mapping, digital twins, etc.

It can be seen from the definition of the product digital twin model that: (1) the product digital twin model is a simulation model in which product physical entities are integrated in the information space, a digital file of the entire lifecycle of product physical entities, and the integrated management of the product lifecycle data and full value chain data; (2) the product digital twin model is perfected by continuous data and information interaction with the physical entity of the product; (3) the final representation of the product digital twin models is a complete and accurate digital description of the physical entity of the product; and (4) product digital twin model can be used to simulate, monitor, diagnose, predict, and control the formation process and status of physical entities in a physical environment.

The product digital twin model is far beyond the category of digital prototype (or virtual prototype) and digital product definition. The product digital twin model includes not only the description of the product geometry, function and performance, but also the description of the formation process and state of the whole life cycle, such as product manufacturing or maintenance process. Digital prototype, also called virtual prototype, is a digital description of a mechanical product or a subsystem with independent functions. It not only reflects the geometric properties of the product object, but also reflects the function and performance of the product object in at least one domain. Digital prototype is formed in the stage of product design and can be applied to the whole lifecycle of products, including engineering design, manufacturing, assembly, inspection, sales, use, after-sale, recovery and other links. The definition of digital product refers to the activities of digitizing the function, performance and physical properties of mechanical products. From the connotation of digital prototype (or virtual prototype) and digital product definition, they mainly focus on the description of the product geometry, function and performance in the product design stage, and does not involve the description of the formation process and state of other full life cycle stages such as product manufacturing or maintenance process.

2.3. Basic features of the product digital twin model

The product digital twin model has many characteristics including: virtuality, uniqueness, multiphysical, multiscale, hierarchical, integrated, dynamic, super-realistic, computability, probability and multidisciplinary.

1. Virtuality: the product digital twin model is a physical product in digital mapping model, information space is a virtual model, belonging to the information space (or virtual space) and does not belong to the physical space.
2. Uniqueness: a physical product corresponds to a product digital twin model.
3. Multiphysical: the product digital twin model is based on the physical properties of physical product digital mapping model; It is not only necessary to describe the geometric properties of the physical product (such as shape, size, tolerance, etc.), but also to describe the various physical properties of the physical product, including structural dynamics models, thermodynamic models, stress analysis models, fatigue damage models, and material properties of product composition materials (such as stiffness, strength, hardness, and fatigue strength).
4. Multiscale: the product digital twin model not only describes the macroscopic properties of the physical product, such as geometric dimensions, but also the microscopic properties of the physical product, such as the microstructure of the material, the surface roughness and so on.
5. Hierarchical: the different components, parts, etc. that make up the final product can all have their corresponding digital twin models. For example, the aircraft digital twin model includes the rack digital twin model, the flight control system digital twin model, the propulsion control system digital twin model, etc., which is conducive to hierarchical and detailed management of product data and product models, and the progressive realization of the product digital twin model.
6. Integrated: the product digital twin model is a multiscale and multilevel integrated model of multiple physical structure models, geometric models, and material models, which is conducive to the rapid simulation and analysis of the product's structural and mechanical properties.
7. Dynamic: the product digital twin model will constantly change and improve through the continuous interaction with the product entity during various stages of the whole lifecycle; for example, product manufacturing data (such as test data, the progress data) will be reflected in the digital twin model of the virtual space, and at the same time, based on the digital twin model, can realize the real-time, dynamic and visual monitoring of the manufacturing state and process of the product.
8. Super-realistic: the product digital twin model and the physical product are basically identical in appearance, content, and nature, with high degree of actuality, and can accurately reflect the real state of the physical product.
9. Computability: based on the product digital twin model, simulations, calculations and analysis can be used to simulate and reflect the status and behavior of the corresponding physical product in real time.
10. Probability: the product digital twin model allows computation and simulation using probabilistic statistics.

11. Multidisciplinary: the product digital twin model involves the intersection and fusion of multiple disciplines such as computational science, information science, mechanical engineering, electronic science, physics, etc., and has multidiscipline.

3. Relationship between digital twin and digital thread

Digital thread was first proposed by Lockheed Martin of the United States. In the production of F-35, they directly input MBD data into computer numerical control machine tools to process components, or complete the laying of composite materials through the programming system, and called this new working mode "digital thread." The "digital thread" saves 6000 sets of tooling for the three configurations of the F-35. It also eliminates the time required for the management of these tooling and the configuration of the parts, as well as the time it takes to distribute the tooling and load it onto the machine.

The US Department of Defense uses digital thread as the most important basic technology for digital manufacturing. Boeing has already been pushing forward the digital mainline technology of single data source product to interact. The National Center for Manufacturing Sciences NCMS (the National Center for Manufacturing Sciences) has confirmed that digital manufacturing as "one of America's largest and most potential competitive assets" and as a key strategy for the future. The digital main line, as its name suggests, is the main theme of digital manufacturing undoubtedly. The Industrial Internet Alliance does not hesitate to use the digital thread as key technology that the Industrial Internet Alliance needs to focus on. It is no exaggeration to say that the digital thread is the key to the revitalization of the US manufacturing industry. The digital thread is a strong connection between OEMs (manufacturers), operation and maintenance service providers, suppliers and end users. The background generated by the digital thread is based on the "model-centered," and the model here is provided with complete information rich, established in accordance with uniform open standards, norms and semantics of digital model, and it can be read steadily and unambiguously by the machine (or system). On this basis, the digital thread integrates and drives the modern product design, manufacture and guarantee process, allowing the models of each link to timely synchronize and communicate the key data bidirectionally. The principle is shown in **Figure 2**. It can be seen in the design and production process that the parameter simulation model is transmitted to the product definition of full three-dimensional geometric model and digital line processing into the real physical product, and then reflected in the product definition model by online digital detection/measurement system. Then, it is fed back to the simulation analysis model to realize a two-way data transmission process. The core of the digital line is how to build a collaborative environment that covers the entire process of product development, so that the unified model can realize bidirectional flow, reuse, and continuous enrichment of data during all phases of product development.

The "Industry 4.0" terminology team defines digital thread as: digital data stream of product design, manufacturing, and assurance, which is constructed with advanced modeling and simulation tools and cover the whole life cycle and value chain of product, and are integrated and driven by a unified model from all aspects of basic materials, design, process, manufacturing, and use and maintenance.

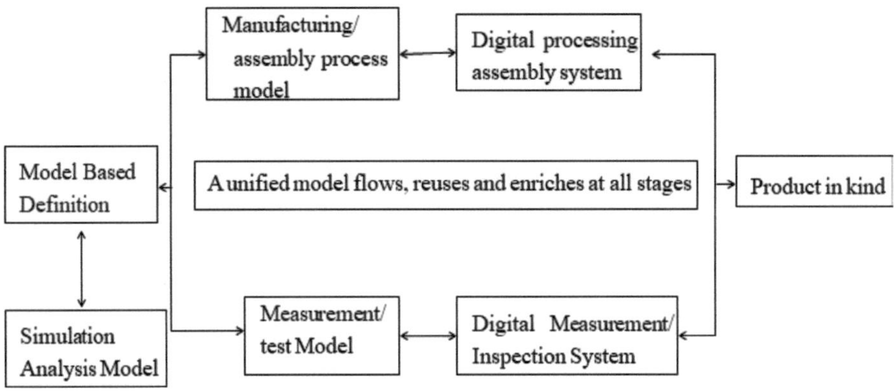

Figure 2. Principle of digital thread.

It can be found that digital twin is a concept that is related to and differentiated from digital thread. Digital twin is a digital representation of a physical product, so that we can see on this digital product what the actual physical product may be and related technology that includes augmented reality and virtual reality. Digital thread in the process of design and production, the parameters of the simulation analysis model can be passed to the product's full three-dimensional geometric model, and then transmitted to the digital production line to be processed into a real physical product, and then reflected in the product definition model through the online digital detection/measurement system, and it is fed back to the simulation analysis model, so that the current and future functions and performance of the dynamic and real-time evaluation system can be realized.

In short, digital thread runs through the entire product lifecycle, especially from the seamless integration of product design, production and operation, while digital twin, more like the concept of intelligent products, emphasizes feedback from product operation and maintenance to product design. It is the digital shadow of physical products. Through integration with external sensors, it reflects all the characteristics of objects from micro to macro and shows the evolution process of product life cycle. Of course, not only products but also production systems (production equipment and production lines) and systems in use and maintenance should be built as needed.

4. Implementation of digital twin technology

4.1. Product design stage

Product digital twin is a hyperrealistic dynamic model of physical products in virtual space. In order to achieve product digital twin, we must first have a natural (easy to understand), accurate, efficient digital expression method of data definition and transmission. The method supports all stages of product life cycle, including product design, process design, processing, assembly, use and maintenance. The model-based definition (MBD) technology that has emerged in recent years is an effective way to solve this problem, and therefore it has become one of the important means for achieving product digital twin. MBD refers to a digital definition method that attaches all relevant product design definitions, process descriptions, attributes, and management information to the product's three-dimensional model. MBD technology enables product definition data to drive all aspects of the entire manufacturing process, fully modeling the concept of parallel collaborative design of products and the idea of a single data source, which is also one of the essence of digital twin. The product definition model mainly includes two kinds of data: one is geometric information, that is, the product design model; and the other is nongeometric information, stored in the specification tree, and the PDM software supporting the 3D design software which is responsible for storing and managing the data [16].

Secondly, after the product definition based on the 3D model is realized, the process design, tooling design, production manufacturing process and even product function testing and verification process simulation and optimization need to be performed based on the model. In order to ensure the accuracy of simulation and optimization results, at least the following three points must be guaranteed:

1. High accuracy and hyperrealism of product virtual model: product modeling not only needs to pay attention to geometric feature information (shape, size and tolerance) but also the physical properties of the product model (such as stress analysis model, dynamic model, and thermodynamics model, and material stiffness, plasticity, flexibility, elasticity, fatigues strength, etc.). Through the use of artificial intelligence, machine learning and other methods, based on the historical data of similar product groups to achieve continuous optimization of existing models, the product virtual model is closer to the functions and characteristics of real-world physical products.

2. Accuracy and instantaneity of simulation: advanced simulation platform and simulation software can be used, such as commercial simulation software Ansys, Abaqus, etc.

3. Model light-weighting: model light-weighting is the key technology for achieving digital twin. First of all, the lightweight technology of the model greatly reduces the storage size of the model, so that the geometric information, feature information and attribute information needed for product process design and simulation can be directly extracted from the 3D model without any unnecessary redundant information. Second, the lightweight model makes it possible to visualize product simulations, simulate complex systems, simulate production lines, and simulate products based on instant data. Finally, the lightweight model reduces the time, cost, and speed of information transfer between systems, facilitates end-to-end integration of value chains, information sharing between upstream and downstream companies in supplies chains, business process integration, and collaborative product design and development.

4.2. Production stage

The evolution and improvement of product digital twin is through constant interaction with product entities. In the manufacturing phase, the physical real world delivers production test data (such as test data, schedule data, and logistics data) to virtual products in the virtual world and displays them instantly. Product's model-based production test data monitoring and production process monitoring are realized (including the comparison of the design value and the measured value, the comparison of the actual used material characteristics with the design material characteristics, the comparison of the planned completion schedule and the actual completion schedule, etc.). In addition, based on the production of measured data, through the intelligent forecasting and analysis of logistics and schedules, we can realize the prediction and analysis of quality, manufacturing resources, and production schedules. At the same time, the intelligent decision module formulates a corresponding solution to the entity product based on the results of the prediction and analysis, so as to achieve dynamic control and optimization of the entity product, and achieve the purpose of virtual integration and virtual control.

Therefore, how to achieve real-time accurate multisource heterogeneous data collection, effective information extraction, and reliable transmission in a complex and dynamic physical space are prerequisites for achieving digital twin. In recent years, the rapid development of technologies, such as Internet of Things, sensor networks, industrial Internet, and semantic analysis and identification, has provided a practical and feasible solution. In addition, artificial intelligence, machine learning, and data are used to demonstrate the role of digital twin in product data integration demonstration, product production progress monitoring, product quality monitoring, intelligent analysis and decision-making (such as product quality analysis and forecasting, dynamic scheduling and optimization). The rapid development of technology such as mining and high-performance computing has provided important technical support for this purpose. Since the assembly line is the carrier for product assembly, the architecture also considers digital hygienic production and assembly line digital twin. The framework mainly includes three parts:

1. Real-time collection of dynamic data in physical space: the dynamic data generated during the assembly process of the product can be divided into production personnel data, instrument and equipment data, tooling tool data, production logistics data, production progress data, production quality data, and actual work hour data. There are eight categories of data for reverse problems. First of all, for the manufacturing resources (production personnel, equipment, tooling, materials, AGV, pallets), combined with the characteristics and needs of the production site, the use of barcode technology, RFID, sensors and other Internet of Things technology, manufacturing resource information identification, the manufacturing process awareness information collection point is designed and a manufacturing object connection network is constructed in the production workshop to realize the real-time perception of manufacturing resources. The production personnel data, instrument and equipment data, tooling data, production logistics data and other manufacturing resources-related data are classified as real-time sensing data; the production progress data, actual work hour data, production quality data, and reverse problem data are classified as process data. Real-time sensing data collection will promote the production of process data. In addition, for the abovementioned large number of multisource, heterogeneous production data, on the basis of predefined manufacturing information processing and extraction rules, the multisource manufacturing information relationship is defined, data identification and cleaning are performed, and data is finally standardized and packaged and formed a unified data service and published it externally.

2. The digital twin evolution of virtual space: by using the unified data service to drive the three-dimensional virtual model of assembly line and the three-dimensional model of product, the product digital twin instances and the assembly line digital twin instances are generated and updated continuously. Assembling line digital twin and product digital twin instances are associated with real-world assembly lines and physical products, and the data exchange between each other is achieved through a unified database in virtual space.

3. Status monitoring and process optimization feedback control based on digital twin: real-time monitoring, correction and optimization of product production process, assembly line and assembly station through historical data of assembly line, excavation of product history data, and assembly process evaluation technology through the comparison of real-time data and design data and planning data, the comparison of product technology status and quality characteristics, real-time monitoring, quality forecasting and analysis, advance warning, and production scheduling optimization are realized, so as to achieve closed loop feedback of product production process and bidirectional connection between control and virtual reality. Specific functions include real-time monitoring of product quality, product quality analysis and optimization, real-time monitoring of production lines, real-time monitoring of manufacturing resources, optimization of production scheduling, and optimization of material distribution.

4.3. Product service stage

During the product service (product use and maintenance) stage, the status of the product still needs to be tracked and monitored in real time, including the physical space location, external environment, quality status, usage status, technology, and functional status of the product. The actual status, real-time data, use and maintenance of recorded data predict and analyze the health, life, function and performance of the product, and provide early warning of product quality issues. At the same time, when the product fails and has quality problems, it can realize rapid positioning of product physical location, fault and quality problem records, parts replacement, product maintenance, product upgrade and even scrapping and decommissioning.

On the one hand, in the physical space, using the Internet of Things, sensor technology, mobile Internet technology, the measured data related to physical products (the latest sensor data, location data, external environment sensing data, etc.), product usage data and maintenance data are mapped to the product digital twin in the virtual space.

On the other hand, in the virtual space, the model visualization technology is used to realize the real-time monitoring of the physical product usage process; combining with historical data, historical maintenance data and related historical data of the same type of products, the continuous optimization of product model, structure analysis model, thermodynamic model, product failure and life prediction and analysis model is realized by using machine learning data mining methods and optimization algorithms; it makes the product digital twinning and prediction analysis model more accurate, and the simulation prediction results more in line with the actual situation. For physical products that have experienced faults and quality problems, traceability and simulation techniques are used to quickly locate quality problems, cause analysis, solution generation, and feasibility verification. Finally, the final results generated are fed back to the physical space to guide the product quality troubleshooting and tracing. Similar to the product manufacturing process, the implementation framework of digital twin in the process of product service mainly includes three parts: data collection in physical space, digital twin evolution in virtual space, and state monitoring and optimization control based on digital twin.

5. Research progress and application of digital twin technology

5.1. Research progress on digital twin

The concept of digital twin was first proposed by professor Grieve in 2003 at the University of Michigan's product lifecycle management course and was defined as a three-dimensional model including physical products and virtual products and the connection between the two. However, due to technical and cognitive limitations at that time, the concept of digital twin was not taken seriously. It was not until 2011 that the US Air Force Research Laboratory and NASA jointly proposed the construction of a digital twin for future aircrafts, and defined digital twin as a highly integrated multiphysical field, multiscale and multiprobability simulation model for aircrafts or systems. It was able to reflect the function, real-time state and evolution trend of the entities corresponding to the model by using physical model, sensor data and historical data, etc. Then, digital twin really attracted attention. Some scholars supplemented and perfected it on the basis of NASA's concept. For example, Gabor and others suggested that digital twin should also include expert knowledge to realize accurate simulation. Rios and others believed that digital twin was not only for aircrafts [17, 18].

In the process of continuous improvement and development of the concept of digital twin, academia has mainly carried out relevant research on modeling, information physics integration, interaction and collaboration, and service application of digital twin.

Some research has been carried out on the framework and modeling process of digital twin modeling in modeling, but there is still no consistent conclusion. Some progress has been made in modeling-related theories, including physical behavior research, nondestructive material measurement technology, quantitative error and confidence evaluation research. These auxiliary technologies will help to determine model parameters, construct behavior constraints, and verify model accuracy.

In the aspect of information physics fusion, there are only preliminary studies on the dimensionality reduction and integration of sensor data and manufacturing data in the aspect of digital twin information physics fusion, while the research on the theory and technology of digital twin information physics fusion is still blank. In order to solve this difficult problem, professor Tao Fei decomposed and refined the scientific problem of information physics fusion into four different dimensions of fusion: physical fusion, model fusion, data fusion, and service fusion in 2017, and designed the corresponding system implementation reference framework. Combined with the theory of digital twin technology and manufacturing service, this chapter systematically studies and discusses four key scientific issues of physical integration, model integration, data integration and service integration, and extracts and summarizes the corresponding basic theories and key technologies. His related work provides some theoretical and technical references for relevant scholars to carry out theoretical and technological research on the physical integration of digital twin information and for enterprises to build and practice the concept of digital twin.

The research on real-time acquisition theory of production data and man-machine interaction that has been carried out in the field of interaction and collaboration is helpful to realize the interaction and collaboration between the physical world and the virtual world. However, there are few related researches on interaction and collaboration between machines and services at present.

In service application, some research has been carried out on service application of digital twin in fatigue damage prediction, structural damage monitoring, real-time running states detection, faults location, etc.; however, there are still many problems to be solved in realizing service integration and coordination. From the above analysis, it can be seen that the research on the related theories of digital twin is still in its infancy. In order to promote the application of digital twin to the ground, it needs to be systematically and deeply studied in the aspects of digital twin modeling, information physics fusion, exchange and cooperation.

5.2. Application of digital twin technology

5.2.1. Product design based on digital twin

Product design refers to the work process of providing all the solutions needed for product production through research, analysis and design according to user requirements. The product design based on digital twin refers to the synergy of existing physical products and virtual products in the design driven by the digital data generated by the products, and continuously discovers new, unique and valuable product concepts and transforms them into detailed products. The design plan continuously reduces the inconsistency between the actual behavior of the product and the expected behavior of the design. The product design based on digital twin emphasizes the overall improvement of design quality and efficiency through the integration of virtual and real life cycles and the establishment of virtual simulation models with super reality.

5.2.2. Virtual prototype based on digital twin

Virtual prototype is a digital model built into the digital world that reflects the authenticity of a physical prototype, through multi-domain comprehensive simulation and equipment performance attenuation simulation, the performance of the equipment can be tested and evaluated before the physical prototype is manufactured, and the design defects can be improved to shorten the design improvement period. The virtual prototype based on digital twin is based on the comprehensive and realistic description of the mechanical system, electrical system and hydraulic multidomain system of the equipment. It has the ability to map the life cycle of the physical equipment, thus designing the equipment, and predictive maintenance provides powerful analytical decision support.

5.2.3. Workshop rapid design based on digital twin

The workshop rapid design based on digital twin adopts the idea of "information physics fusion," which completes the digitization of physical equipment, the scripting of motion process, the integration of the whole system, the synchronization of control commands, and the parallelization of on-site information to form a complete line of execution engine. Through the physical equipment and the corresponding virtual model for virtual and real interaction, instruction and information synchronization, a rapid design, planning, assembly and testing platform for the workshop supporting physical equipment connection is formed. The platform: (1) uses the 3D engine designs and builds a special model library, combining workshop area, capacity requirements, equipment selection, and construction of virtual 3D model of workshop, which can quickly complete the workshop layout design; (2) prepares action scripts for heterogeneous devices, develops response programs, builds virtual control networks, implements near-physical simulation of virtual full-line machining movements, and predicts, evaluates and optimizes based on actual data; and (3) can test the consistence of the distributed integration equipment and the whole line movement, the internal control logic, the instruction and information downlink channel, the job cycle synchronization and so on, and optimize the workshop design based on the virtual reality fusion data.

5.2.4. Process planning based on digital twin

Process planning is the technical document of the product manufacturing process and operation method; it is a disciplined document that all production personnel should strictly and conscientiously implement; and it is the basis for product production preparation, production scheduling, worker operation and quality inspection. Digital twin-driven process planning refers to the realization of process design and continuous optimization for production sites by establishing virtual simulation models such as products, resources and process flows with super reality, and virtual and real mapping of full factor and full process. In the process design mode of digital twin driving, the simulation model of virtual space and the entity of physical space are mapped to each other to form an iterative collaborative optimization mechanism of virtual and real symbiosis.

5.2.5. Workshop production scheduling optimization based on digital twin

Production scheduling is the nerve center of decision-making optimization, process control and performance improvement in the production workshop, and it is the operation pillar of orderly, stable, balanced economy, and is agile and efficient in the production workshop. The digital twin-driven scheduling mode is a new scheduling mechanism of virtual-real response, virtual-real interaction, virtual-control-real, iterative optimization, which is supported by the digital twin system, through the virtual-real mapping and interaction fusion of all elements, all data, all models and all spaces, and realizes cooperative matching and continuous optimization of the "workpiece-machine-constraint-goal" scheduling requirements. Under the digital twin-driven scheduling model, the scheduling elements are mapped to each other in the physical workshop and virtual workshop, forming the co-optimization network of virtual reality co-existence. Physical workshop actively perceives production status. Virtual workshop can analyze scheduling status, adjust scheduling scheme and evaluate scheduling decision through self-organization, self-learning and self-simulation. It can quickly determine abnormal range, respond quickly and make intelligent decision. It has better adaptability to change, disturbance response ability and abnormal resolution ability.

5.2.6. Production logistics accurate distribution based on digital twin

Production logistics including enterprise internal logistics and enterprise external logistics between businesses are to guarantee the normal production and are the key of high production efficiency, and they reduce the product cost. Digital twin production logistics is under the twin data-driven, and through the actual physical entities and the virtual model mapping, real-time interaction and closed loop control, realize the task of production logistics combinatorial optimization, transportation route planning, transportation process control in the physical world and information world and overlapping generations between the top logistics services system, so as to achieve production logistics seamlessly and is an intelligent of a new kind of production logistics operation mode.

5.2.7. Intelligent control of workshop equipment based on digital twin

The control system of workshop equipment is the brain of workshop equipment. The correctness of its control function and control strategy directly affects the function and performance of workshop equipment. The control advantages of digital twin are as follows: (1) in the stage of equipment design, the design of the control system is matched based on the digital twin virtual reality synchronization, so that the control system and the physical equipment are fused earlier and match, and the burden of the real machine is lightening; (2) in the commissioning stage, the overall matching of control system and equipment is further promoted, the design defects are improved, and the design redundancy is reduced; and (3) in the running stage, the control feedback information is no longer a relative independent parameter, but the physical real time state of the digital twin, which can provide objective and effective data support for the autonomous decision of the algorithm.

5.2.8. Man-machine interaction based on digital twin

By human-computer interaction, the flexibility of the machine can be improved and the workload of manual work can be reduced. The workshop man-machine interaction based on digital twin refers to the construction of a digital twin virtual workshop which is fully mapped to the actual physical workshops. Through high speed and reliable communication technology, the robot can quickly adjust the work plan by identifying the workers' instructions through touch, gesture, or sound, so as to make it possible to cooperate with the workers. Industry action and update the manufacturing process of virtual workshop in real time.

5.2.9. Assembly based on digital twin

The assembly of complex products is the final stage and key link of the realization of product function and performance. It is an important factor affecting the quality and performance of complex products. The quality of assembly determines the final quality of complex products to a great extent. The assembly process of the digital twin drive will be based on the integration of the Internet of Things of all equipment, the integration of the physical world of the assembly process and the information world, and the precise control of the parts, equipment and assembly process through the intelligent software services platform and tools, and the unified and efficient control of the assembly process of complex products. The self-organization, adaptation and dynamic response of the product assembly system are realized.

5.2.10. Testing/detection based on digital twin

The digital twin drive test/detection is to build a high-fidelity test system and a virtual model of the measured object in the virtual space. With the help of test data real-time transmission and test instruction execution technology, the multi-discipline/multi-scale/multi-physical properties of the physical object under test and the virtual object under test are driven by historical data and real-time data to achieve high-fidelity simulation and interaction. It thus intuitively and comprehensively reflects the full life cycle state of the production process and effectively supports scientific decision-making based on data and knowledge. The digital twin-driven test/detection process includes knowledge modeling, system design, system construction, and full lifecycle management and autonomous decision-making of system, object, and process state data.

5.2.11. Manufacturing energy management based on digital twin

The management of manufacturing energy consumption refers to monitoring, analyzing, controlling and optimizing the energy consumption of water, electricity, gas, heat and raw materials in the manufacturing process, while ensuring the performance of the manufacturing system and the economic benefit of the enterprise, so as to realize the fine management of energy consumption, achieve energy saving and reduce the cost of the manufacturing enterprise, and maintain the enterprise, the purpose of the competitiveness of the industry.

5.2.12. Product quality analysis and traceability based on digital twin

Product quality analysis and tracing refers to the design of the correct and reasonable manufacturing process, at the same time, the processing precision, stress and other factors in the production process are comprehensively considered to realize the analysis of the quality of the product. In the case of quality problems, it can trace all the links in the processing and find out the reasons, thus improving the processing technology and control, processing quality.

5.2.13. Fault prediction and health management based on digital twin

Prognostics and Health Management (PHM) use various sensors and data processing methods to evaluate the health status of the equipment, and predict the equipment failure and residual life, so as to transform the traditional post maintenance into pre service maintenance.

5.2.14. Product-service system based on digital twin

Product-service system (PSS) is a value providing system that provides a combination of different "physical products and services" to consumers, including a product oriented PSS, a use oriented PSS, and a result oriented PSS. Under the support of digital twin, PSS is based on the digital twin, through the intelligent analysis and decision-making of different "physical products and service" combination, rapid personalized product service configuration and service process experience and rapid supply, and the use of the virtual and real synchronization among elements to realize the optimal allocation and integration of resources. The PSS based on the digital twin model makes full use of the digital and information system to effectively support the intelligent decision-making, rapid supply, intelligent service, value and environment analysis of the life cycle of complex products and services.

6. Conclusions

The digital twin technology can not only make use of the theories and knowledge of human beings to establish virtual models, but also make use of the simulation technology of virtual models to explore and predict the unknown world, and find better ways, constantly stimulate the creative thinking of human beings, and continue to pursue the optimization and progress, which are the innovation of the current manufacturing industry. This chapter mainly summarizes the definition, connotation and implementation methods of digital twin technology.

Acknowledgements

During the writing of this chapter, doctoral candidate Yan He, master graduates student Zijian Zhang, Yujie Yuan, Yu-hu Li, Yansong Liu, Hao-wei He, Wangxing Yan and Tengfei Wang of CAD/CAM Engineering Technology Research Center, School of Mechanical Engineering, North University of China and Shanxi Crane Digital Design Engineering Technology Research Center provided their help. They did a lot of work for this chapter, and a special thanks for their efforts.

Author details

Zongyan Wang

Address all correspondence to: iamwangzongyan@sina.com

School of Mechanical Engineering, North University of China, Taiyuan, China

References

[1] Rosen R, Von Wichert G, Lo G, et al. About the importance of autonomy and digital twins for the future of manufacturing. IFAC-PapersOnLine. 2015;**48**(3):567-572

[2] Grieves M. Digital twin: Manufacturing excellence through virtual factory replication. Digital Twin White Paper. 2014. http://innovate.fit.edu/plm/documents/doc_mgr/912/1411.0_Digital_Twin_White_Paper_Dr_Grieves.pdf

[3] Grieves M. Product lifecycle management: The new paradigm for enterprises. International Journal of Product Development. 2005;**2**(1/2):71-84

[4] Grieves M. Product Lifecycle Management: Driving the Next Generation of Lean Thinking. New York, NY, USA: McGraw-Hill; 2006

[5] Lee E. Computing Foundations and Practice for Cyber-Physical Systems: A Preliminary Report, Technical Report UCB/EECS-2007-72. USA: University of California; 2007

[6] Grieves M. Virtually Perfect: Driving Innovative and Lean Products Through Product Lifecycle Management. Cocoa Beach, FL, USA: Space Coast Press; 2011

[7] Tuegel EJ, Ingraffea AR, Eason TG, et al. Reengineering aircraft structural life prediction using a digital twin. International Journal of Aerospace Engineering. 2011;**2011**:14. Article ID: 154798. http://dx.doi.org/10.1155/2011/154798

[8] Eric Tuegel. The Airframe Digital Twin: Some Challenges to Realization. https://arc.aiaa.org/doi/abs/10.2514/6.2012-1812

[9] Glaessgen EH, Stargel DS. The Digital Twin Paradigm for Future NASA and U.S. Air Force Vehicles. https://ntrs.nasa.gov/archive/nasa/casi.ntrs.nasa.gov/20120008178.pdf

[10] Shafto M, Conroy M, Doyle R, et al. Modeling, Simulation, Information Technology & Processing Roadmap. Washington, DC, USA: NASA; 2012

[11] Lee J, Bagheri B, Kao HA. A Cyber-Physical Systems Architecture for Industry 4.0-based Manufacturing Systems. Manufacturing Letters. January 2015;**3**:18-23

[12] Warwick G. GE advances analytical maintenance with digital twins. Aviation Week & Space Technology. 2015

[13] Fei T, Meng Z, Jiangfeng C, et al. Digital twin workshop: A new paradigm for future workshop. Computer Integrated Manufacturing Systems. 2017;**23**(1):1-9

[14] Cunbo Z, Jianhua L, Hui X, et al. Connotation, architecture and trends of product digital twin. Computer Integrated Manufacturing Systems. 2017;**23**(04):753-768

[15] Fei T, Ying C, Jiangfeng C, et al. Theories and technologies for cyber-physical fusion in digital twin shop-floor. Computer Integrated Manufacturing Systems. 2017;**23**(08):1603-1611

[16] Yong Y, Shengting F, Guanwei P, et al. Study on application of digital twin model in product configuration management. Aeronautical Manufacturing Technology. 2017; **526**(7):41-45

[17] Bohu L, Lin Z, Shilong W, Fei T, et al. Cloud manufacturing: A new service-oriented networked manufacturing mode. Computer Integrated Manufacturing Systems. 2010; **16**(1):1-7

[18] Fei T, Weiran L, Jianhua L, et al. Digital twin and its application exploration. Computer Integrated Manufacturing Systems. 2018;**24**(1):1-18

The Role of Spin-Off Companies in the Technology Transfer and IS Management Potential in Developing a Sharing Economy

Mária Pomffyová, Mária Rostašová and
Vladimír Krajčík

Additional information is available at the end of the chapter

http://dx.doi.org/10.5772/intechopen.81441

Abstract

The development of all dimensions of digitization, in particular the global Internet penetration into most business areas, creates a better condition for sharing technology. Start-up and spin-off companies, supported by university centres, directly support the transfer of the latest technology into practice. Using the model of economic management, the effectiveness of such economics will be researched. Practices and recommendations that should eliminate barriers to successful spin-offs and start-ups as well as SMEs doing business in competitive area will also be proposed. The assessment of the views of 189 managers of Slovak SMEs and 26 managers of IT companies creates the opportunity to assess the extent to which companies want to apply innovation opportunities and IT potential in practice as a driver of changes. These results will be compared with the results of similar research in the EU. The verification of the mutual dependencies of selected indicators using statistical methods and validation tools that are part of the SPSS program will be provided. Based on the synthesis of all knowledge and experience, it will be suggested to managers the ways how to apply technology transfer and regulatory treatment of sharing economy in practice as effectively as possible.

Keywords: sharing economy, start-up and spin-off company, transfer of information and communication technology, innovations, online platform

1. Introduction

The advantages lie in their current technological background and scientific innovation support. In this chapter, we will evaluate the conditions for the development of the sharing economy from the point of view of management of technology transfer. The development of sharing economy depends on the possibility of enhancing the sharing of different surplus services, products, or solutions between providers and consumers by using nontraditional forms of exchange using the latest information and communication technologies (ICTs), as well as Internet as a space for interconnection with various information platforms or other intelligent devices. People are closer to each other due to professional Internet applications. The better the conditions for sharing technology, the more effective the levels of sharing. The basic assumptions are created using the development of all dimensions of digitization, in particular the dimension of global Internet expansion, its penetration into most business, public and private activities, or modifying expansion. New solutions should be available from different environments using unified solutions for different applications. However, their development and application into the practice will not succeed without the need, willingness, and ability of people to offer, develop, and exploit their potential, accessed through the necessary services or products using simple and unpretentious solutions. Therefore, nowadays, more attention should be paid to research the conditions for setting up and supporting the business doing of spin-off companies in such regions [1]. In this chapter, we also will evaluate the conditions for the development of the sharing economy from the point of view of IT transfer as a tool for sharing products or services in that ways of doing business regarding potential of such regions. We will also examine the possibilities of economic model for management of environment for sharing technologies.

2. Theoretical background of sharing economy and the role of electronic business support

If we want to formulate principles of sharing economy as an alternative model of traditional economy, we can find many different approaches to their evaluation. In addition, some authors did not agree with the notion of "sharing economy" and introduce other concepts, in identical or similar terms.

2.1. Approaches to evaluation of the sharing economy principles

The sharing economy can be defined as an economy that is based on the principle of accessing services and goods, which are owned by another person, but they are surplus and are available, and it is possible to provide them to another user (either because their total capacity is not being used or there exists the intention to use them just in such way), depending on the creation of conditions for a sharing economy. The principles of "sharing and sharing the economy" lie in the existence of a certain good (or potential of free service capacity). On the other hand, the existence of an entity (e.g. a natural person) interested in using such good or service, even for a fee, is assumed.

The definition of the sharing economy varies according to the approaches to assessing the principles of its functioning. Many authors argue that the model of a sharing economy is a competing model for a classic system of product and service delivery.

Some authors point out that consumers buy the traditional market products (which they own then) and services, while in the case of a sharing economy, suppliers temporarily share their goods and services with customers either for free or for financial or nonfinancial remuneration [2].

Eckhardt and Bardhi [3] state that the "access economy" is a more precise term for a sharing economy. "Sharing is perceived as a form of social exchange that takes place among people (e.g. family, neighbors, friends) without any profit." They say the advantage of an "access economy" lies in two key success factors:

- Competition between the providers will not depend on whether the provider offers greater social interactions and community ownership, but acceptable comfort at a lower cost.
- Consumers will be driven more by their benefits and experiences than marketing oriented to communicate brand and its ownership [4]; if they can try, for example, to taste more identities directly, they will feel better in their decision-making.

These statements confirm the benefits of sharing economy, which offers the opportunity to assess the quality of products offered within the EU, without being a restriction on the place of purchase or the place of origin of the product, respecting the importance and role of marketing.

"Circular economy" is an alternative expression for a sharing economy that emphasizes the reuse of product features more than sharing them. At the same time, it points out that reuse has a beneficial effect on the environment, better use of materials, and energy embodied in products [5]. Reuse may be:

- parallel (good may use more people—the term nonlife consumption is usually used in economics in this sense—for example, road use by many drivers at the same time, but also more people transport by the same car);
- serial (for example, you can use more individuals not at the same time but gradually over time—for example, accommodation in a private apartment).

The sharing economy was also evaluated as a "swarm economy" by Rick Falkvinge [6], which is considered as an economy based on a weakening power of large entities. The existence and exploitation of the benefits of electronic communications reduces the need to have only classical deals. "Swarm economy" is based on massive decentralization and overcoming traditional economic models. He also states:

- for contact of the seller and the buyer, an Internet platform is sufficient;
- goods or services are also provided using supplementary platforms;

- it is not the unusual phenomenon that the traditional model of employment will be overcome, one will perform several jobs simultaneously, one or two to subsistence and existence, another for fun;
- customer can listen to the music or watch the movie again at the current time and he/she will not wait for what television or radio will offer him;
- it will probably have to change advertising, the individual will not be dependent on buying everything, and if he will be more in control of the advice of price and price comparators than the advertisement;
- even class distributors will have to change if a number of stone shops now offer goods through e-shops in parallel, perhaps they will have to go to the principles of a shared economy.

In [7], it was stated that the term "sharing economy" was in the year 2015 introduced into the Oxford English Dictionary but draws attention to the fact that a number of other terms are used in the same or similar terms: "peer economy," "collaborative economy," "gig economy," and "on-demand economy." For completeness, let us add that the abovementioned dictionary defines a sharing economy as an "economic system in which assets or services are shared between private individuals, either free of charge or for a fee, usually via the Internet."

The integral part of sharing economy is also a sharing city, where the goal is to "connect forces of technology, citizens, data, and design to the most effective interconnection of people, products, and services" [8]. Their provision must reach a certain level with additional support to be part of the range of services and product consumption. An important role is also played by "e-government." It represents "the use of information technology by public institutions for the exchange of information with citizens, private organizations, and other public institutions in order to increase the efficiency of internal functioning and providing fast, affordable, and high-quality information services" [9]. It is clear that, besides the elements that integrate commercial platforms, like in the case of a sharing city, in a sharing economy, public data and citizens of the cities should also be involved in order to simplify and make the use of public services more efficient. It increases the role of Internet and new technologies used as a support for electronic ways of communication.

2.2. Specific aspects of the sharing economy

From a macroeconomic point of view, the sharing economy does not create a new market; the goal is only to take over a certain part of the existing market, not to generate new customers. It means that it only takes over traditional distribution channels and service providers and provides them with the use of nontraditional ways. People can switch from being only consumers of classically supplied goods or services to those who drive new services enriched with new experiences or events. They can become even designers as we can see in community housing (cohousing, respectively, baugruppe), in multifunctional use of former buildings and spaces, or in sharing cultural activities. To develop a sharing economy, some assumptions should be created. As we can see, for many people, it is more acceptable to have operational access to products or services when they need it, which is a move away from owner rights to user rights [10]. Especially, young people prefer the dynamics of the availability of products and services provided in nontraditional ways, where they prefer something unexpected, unpredictable, or different in comparison with the traditional ways of their availability or consumption. This fact is confirmed by the results of recent studies [11], where about two-thirds of respondents said they are heading toward a less materialistic lifestyle and four-fifths of respondents believe that rent has greater advantages than ownership. It is due to opportunities that the sharing economy allows:

- sharing services and goods between unknown entities, over long distances, outside closed family or neighboring communities, where the reason is also partly the need to discover something new, to gain experience in a nontraditional way, or perhaps as an effort to give up the traditional models to meet their own needs;
- changes in lifestyle, value, especially preferred by younger generations, which is far less associated with property ownership, product gathering, etc., and more preferring to survive something new.

The common practice is that many people increase your own revenue through a business that is a part of the sharing economy. As stated in [11], about 11% of UK residents receive revenue based on running applications that are part of the sharing economy. Of course, there will be a lot of people (the first surveys show about 50%) who will hold the opinion that their stuff will never be shared with others.

The sharing economy can also support the growth of employment and enterprise performance. In the key sectors of sharing economy, such as finance, accommodation, transport, small domestic services, and professional services, the number of operations is expected to increase up to 20 times in comparison with the situation in 2015 [12]. It is due to the opportunity of employing people in productive age, post- as well as preproductive age, as they are not only consumers but also providers or intermediaries of the various products or services. In accordance with nontraditional ways of doing business, the disadvantages associated with sharing economy are also occurring. There are situations where the circumvention of valid business regulation arises and tax payments are not strictly compliant on the part of providers. It is the case that a certain market group operates under certain regulatory conditions (qualifications, insurance, hygiene, etc.), and the other group is not required to formally await their confirmation. It means that conditions for doing business are different, which is why an unequal competitive environment is formed [13]. In order to increase the benefits of the sharing economy, it is necessary to deal with various specific issues such as regulatory treatments and conditions for sharing or easily disposing with nontraditional means of accessing products or services.

2.3. The basic conditions for sharing economy development

In a sharing economy, the common discussion is focused on the problem, how to create a platform for sharing the involved elements and how to support their accessibility.

If companies want to benefit from sharing economy, the basic conditions must be respected. It is necessary to provide the broader accessibility of:

- modern information and communication technologies;
- modern information platforms and Internet;
- large databases accessing, where their disposability allows, for example, sharing of public goods (products and services) created from public funds intended for the public or private institutions (see, e.g., databases arising in connection with the implementation of e-government) as well as for private property;
- technological solutions, as well as smart devices—smart phones that connect people to each other and create an opportunity to access to professional Internet applications much closer;
- availability of the online payment systems.

To do it, in particular, it is very important to bring innovations to traditional areas and sectors as well. It is necessary to deal with the issue of technology transfer and make them accessible to a wide range of users. The creation of a modern online platform and the availability of many different tools should be realized with the most effective combination of new technologies and relevant resources.

As follows, providers and intermediaries, who are involved in sharing economic transactions, gain new experiences and benefits associated with engaging in sharing transactions [14]. In connection with this, there is a new service sector formed, which can also play the role as an employer. The provider sector where operators, who develop, improve, and operate products and services sharing platforms, are included is formed. It is applied to both the new product or service providers and their intermediaries, where:

- on the side of providers, most operators are already trying to push the style of communication so that both partners feel at the same level, which means that the provider does not restrict the applicant,
- on the side of intermediaries, they dispose with new opportunities: to try the role of a taxi driver, accommodation provider, courier, risk borrowing money at higher interest rates, and also attract, seek for unusual leisure time, to enter the liberal working style model, etc.

The B2B platform acquires its importance in such services or products delivering.

2.4. Microeconomic and macroeconomic conditions of sharing economy

The use of information and communication technology brings the macroeconomic advantages of sharing economy that lies in the fact that a new model of interconnection should be observed between supply and demand from the point of view

- of customers, where it is the comfort associated with the fact that information and communication are mediated through the computer screen or mobile display, the tool of connectivity support is continuously updated and has its online support as well as its own transaction support;
- of the information platform, where it is possible to reach a much wider range of customers; the modern communication platform minimizes the level of information asymmetry, where both parties dispose with high-quality information for their transactional decision-making. It should also lead to savings on transaction costs [14], as follows:
- search and information costs—platforms offer fast and often clear and comparative product and service information to help reduce loss due to lack of information or knowledge of the product or service,
- negotiation costs—communication and contract costs,
- implementation costs—shipping agreement, payments, insurance, guarantees, etc.

It is also necessary to address the microeconomic problem where the modern sharing economy combines two unfamiliar partners. In the case of private transactions, the success of cooperation between the partners strongly depends on the degree of fairness and responsible behavior toward each other based on mutual personal relationships. The trust among partners is affected by the degree of personal acquaintance, which is why it is necessary to be respectful and anticipate the behavior of its partner. As business support as well as communication between partners and providers is based on the electronic business support that is fully utilized, it is also possible to exploit the application as a service quality regulator. Many applications offer the ability to perform backup control by which platform providers try to map the dissatisfaction of their clients. In such developed applications, there is a possibility that unsatisfied clients can report their negative experiences, record incidents, and so on. This option addresses problems associated with poor quality of service or product properties that may occur due to the ways of their provision. In such situations, it is important to build on the support of a single electronic platform and to develop an area where electronically supported applications or other means will be easily accessed. Very often, it is necessary to quickly adapt to new technologies and apply them to business practice. In these processes, the importance of the role of spin-offs, start-ups, and also small and medium-sized enterprises is growing because they are more flexible in nontraditional ways of delivering services or products. They are able to quickly respond to market needs, both in terms of developing and using new applications supported by ICT.

3. The ways of new technologies and knowledge transfer into the practice

At present, the university research and scientific departments and their staff, which devote considerable resources and capacity to research and development activities, represent a progress in the field of technological development and delivery. As the research and development sector is not continuously linked to business practice, it is difficult to create the right conditions for technology and knowledge transfer to practice. Many universities are collaborating with practice (e.g. Oxford University Innovation (OUI) [15]), but the problem is the commercialization of technology potential. The goal is to convert technology and knowledge potential into usable products or services. Both sides of this relationship (both, university, enterprise) dispose with a variety of resources: research and human resources, skills, experience, and know-how, as well as intellectual property (IP), certification capacities, permits, production facilities, supply relationships and chains, the possibility of marketing, distribution, etc. In order to achieve successful interconnection between the transfer of technology and knowledge, the following options are available:

- licensing an existing company;
- creating new company and its licensing (spin-off company);
- transformation for further use within the institution;
- moving business to the incubator to establish a start-up firm;
- commercialization of the mutual research results between the research institution and the partner from the commercial sphere (joint venture).

Despite the fact that it is difficult to create the conceptual definition of the establishment of new companies, such as spin-offs and start-ups, their innovative, scientific, and technological potential is a guarantee for future success of such companies. Therefore, we have tried to define spin-offs and start-ups and their role in sharing economy.

3.1. The role, categorization, and typology of spin-offs and start-ups

As stated in [16], the small companies—the so-called spin-offs—play the special role, which consists in research and knowledge transfer. Spin-off is a newly founded company cofounded by a university or research laboratory that owns the licensed technology and applies it to the market with the aim to leverage available academic knowledge for commercialization (firstly established by McQueen and Walmark [17]). This company is a profit-oriented entity. This fact is confirmed by many authors [1, 18–20] who state that spin-off founding helps to create an innovative company that guarantees collaboration with universities and transfer knowledge and technology from universities to practice. The classic university spin-off examples include the Stanford University in Silicon Valley (USA), Google, Hewlett-Packard, Sun Microsystems, Cisco Systems or Silicon Graphics, all established by former university students. Both factors, the founding of spin-off companies and technology licensing, are the basic assumptions for successful technology transfer into real practice.

Certain universities, however, do not use the term spin-off but rather use the term start-up or the term employee's enterprises designed to commercialize IP. Spin-out is also often used instead of spin-off; however, upon closer analysis, both of these terms have a similar meaning.

Many authors have tried to identify common aspects in different spin-off definitions. In [21], spin-offs were divided into two categories:

- spin-offs where the inventor takes part in the business position (active commercialization of proprietary inventions);
- spin-offs where entrepreneurs are not inventors but hold rights for the use of university inventions (position of inventor from the university is replaced by a manager from the practice).

Another important feature is that the university still remains as a co-owner in a given spin-off company but that company can flexibly and freely create your own unique intellectual property (IP). This is confirmed by the authors in [22], who define the university spin-offs as "companies established by one or several university employees who left the university in order to establish legally and technically independent entity which is supported by the university at least in the initial phase of development." The term spin-off is also described as an innovative company established for the use and further development of academic IP. They state that the definition of spin-off within academic conditions reflects various differences in the perception of requirements related to IP commercialization as well as differences in maturity of business environment in different countries. As we have found, in order to define the concept of spin-offs, it is possible to find many different criteria and approaches to their categorization. As stated in [15], it is possible to identify 14 common elements or categorization criteria of different definition and 46 spin-offs categories. In **Table 1**, we assume the results of conceptual categorization of spin-offs based on the ways of establishment of spin-off with relation to the founders or networking creation.

Conceptual framework for new spin-off categories typology		
Criteria, methods, and tools	Research fields	New spin-off categories
• The concept of spin-offs establishment • The ways of spin-off categorization	• Subjects of cooperation • The ways of cooperation • Rate of linkage between spin-offs and employees • Kind of relationship with regard to the founders • Type of subject in the context of networking with external environments • Stage of spin-off	Established in cooperation: • with the university, • by the employee, who remains at the university, • by the employee who leaves the university after its establishment or reduces his/her working time; Established as a result of: • Technology transfer without the involvement of university employees (inventors)—a manager from the practice is hired Connection between the spin-off and • venture capital • business partners (e.g. suppliers) • spin-off and customers • spin-off and competitors

Table 1. Conceptual framework for spin-offs typology.

These fundamentally mentioned categorizations could be extended with new categorization criteria with respect to the industry sector in which the spin-off company operates its business or in terms of company size according to the total number of employees, etc.

Next factor that is more important in terms of properties of spin-offs is the region. The region where the academic institution is located determines the access to the rights of the various parties, also the IP rights, the type and definition of spin-off and start-up companies. We can state that the intellectual property of academic institutions can be viewed as public, private, and social resources. Many spin-off definitions are usually incorporated into internal directions, regulating IP protection and the ways of commercialization at a particular university. At the end, we can summarize that it is more important to pay attention to the regulatory framework, as well as to create acceptable legal or technical conditions for doing business in sharing economy.

4. Actual trends in the area of sharing economy

The European Union (EU) has its own legal system, whose main rules and principles are laid down in the founding treaties [23]. The EU can adopt legislative acts, which member states must comply with and apply. As we have seen, neither the EU nor its own countries have yet defined a single legislative framework that would govern the area of the sharing economy as a whole. In this area, divided into two parts, the platforms of the sharing economy and users of platforms; however, regulatory measures can be found to help producers create better conditions for the operation and support of business activities in the electronic virtual space. In the study [24], the authors summarized a set of measures concerning the legal and regulatory framework for the European Economic Area (EEA) as follows: Directive for E-commerce 2000/31/EC, Services Directive 2006/123/EC, Rights Directive 2011/83/EC, and The Unfair Commercial Practices Directive 2005/29/EC. In 2013, Proposal for a Regulation of the European parliament and for the council laying down measures concerning the European single market for electronic communications and to achieve a Connected Continent, and amending Directives 2002/20/EC, 2002/21/EC and 2002/22/EC and Regulations (EC) No 1211/2009 and (EU) No 531/2012, was introduced [23]. In 2015, the European Commission launched the digital single market (DSM) strategy. According to this strategy, DSM is a comprehensive market in which people and businesses can trade, innovate, and cooperate legally, safely, and at an affordable price. This sector covers areas such as digital marketing, e-commerce, and telecommunications. The single digital market seeks to maintain the rules of fair competition, consumer protection, and the removal of geographic and copyright issues [25].

The DSM is built on three pillars:

- Better access for consumers and businesses to online goods—turning the EU digital world into a single market with the same conditions for sale and purchase.
- An environment in which digital networks and services can be successfully developed—design the rules that respond to the rate of technology and infrastructure development.
- Digitization as a driving force for growth—its role is to ensure that the European economy, industry, and jobs make the conditions for full use of the benefits of digitization.

The DSM strategy, which has been implemented so far, has produced several key and priority legislative proposals addressing e-commerce, copyright, audio-visual and media services, telecommunications services, ePrivacy, harmonization of digital rules, and harmonization of rules on the admission and functioning of value added tax.

In 2016, the European Commission presented also a single market strategy (SMS), where the main objective was to unlock the full potential of the single market for the European Economic Community established by the states of the European Union [26].

There were also presented policy recommendations that should lead to a reduction of barriers that prevent the growth of the sharing economy in the European Union [27].

Regulation (EU) No 1291/2013 of the European Parliament and of the Council of 11 December 2013 establishing a Horizon 2020—the Framework Program for Research and Development—has been adopted to regulate the creation of a framework for research and innovation and Innovation (2014–2020) and repealing Decision No 1982/2006/EC [28].

5. Discussion and methods of research

As Henna Virkkunen, who is responsible for steering plans regarding online platforms and the digital single market modified by the Parliament of the EU [29], states, "the global nature of digital economy increases choices for consumers and boosts price competition." She added that "it offers workers chances to decide how much and where they want to work. Workers should also be able to transfer and accumulate users' electronic ratings and reviews, which constitute their "digital market value" as an equivalent to the traditional market value." In these processes, the question is What is more important?—What is the role of the Internet and traditional business doing? What is the role of their relationship to innovations? and What is the role of spin-off companies in sharing economy?. We tried to find answers to these questions. As the basis for the creation of the conclusions, we provide an evaluation of the research results obtained by performing various economic analyses of indicators of sharing economy, spin-off company functioning, and also the assessment of attitudes of small and medium enterprises to innovations and IT adaptation. These research activities were carried out in the framework of partial research activities of research institutes of the partner universities (University of Žilina, Prague and Institute of Management Systems in Poprad, Matej Bel University, Banska Bystrica) between 2015 and 2017.

6. Research results

Next, we review the current state of the sharing economy and predict its future development. We analyze the available indicators and propose suitable recommendations.

6.1. Revision of the share indicators estimation of sharing economy

An analysis of the amount of unregistered entrepreneurs from the small and microenterprises sector is provided. In the revision of the estimates, their share in the labor market is estimated comparing the outputs of the labor sample survey based on data from the Czech Statistical Office and the Labor Office statistics. The aim of this analysis is to review the ways in which the share of producers deliberately not registered on the labor market is estimated and to assess their verity. The evaluation and comparison of the variety of indicators of sharing economy were provided by using the statistical methods and validation tools that are part of the SPSS program.

As outlined in [13], if we want to identify the database for the analysis of the sharing economy, it is necessary to build on the revisions of national accounts statements[1]. This document describes the impact of changes in methods and changes in data sources within the so-called "major revision of annual national accounts." In the context of revisions, the search and identification of leakages in the form of so-called gray economy. Although the sharing economy is not exactly the same as the gray economy, these two sets overlap in many ways.

It is recommended to follow the estimates of:

- the impact of accommodation services on GDP in the form of dwelling services—imputed rent. In this case, it is necessary to use the stratification method of the imputed rent estimation rather than the unit cost method (UCM). The size of the imputed rent according to (a) the size of the municipality, (b) the type of the building, (c) the size of the apartment, and (d) the equipment is determined;
- producers deliberately not registering—The estimated values are obtained using the analysis of data collection obtained from the pilot study "Exhaustiveness of Czech National Accounts"—from the mutual project of Eurostat and the Czech Statistical Office. The calculation of the values of the indicators is done in two steps (using the old method). First, the percentages obtained by the industry sector are calculated. Secondly, the amount of unofficial unemployment on the basis of the labor market disparity (according to the average labor productivity in the sector) is estimated. In 2010, the estimated value of the variance was 2.8% (employees in full-time equivalents) in CZK 78.513 million, in Euro 2.606 million.

The results of the new method are also obtained in two phases. First, a comparison of generic productivity with productivity that businesses achieve by using credibly leading accounting is provided. Secondly, the estimated value is calculated according to labor market disparities.

The results of the second method are surprising. The deviation is 35.5% (CZK 103.689 million, in Euro 3.442 million €) for employers and 13.4% (CZK 110.846 million, in Euro 3.679 million €) for self-employed persons. As pointed out, the new method allows for more accurate estimates of differences in monitored parameters.

[1] This basic revision was published by the Czech Statistical Office on 30 September 2011 and preceded the main revision in 2014. The main reasons for the revisions were to ensure greater comparability of macro-aggregates in the Czech Republic with respect to the transition to NACE classification. The most important conclusion is the underestimation of the performance of the Czech economy (2–3% absolute in GDP) in the past.

6.2. An analysis of supply and demand in a sharing economy

If we want to assess the situation in terms of demand and supply in the sharing economy, what services or products are preferred, it should be realized using the survey analysis of respondents' opinions that are involved in providing the related services or products.

The aim of the survey should be to find out what is the real state and what is the dynamics of changes in supply and demand in the sharing of the economy by sector, including an estimate of the size of individual markets. On the basis of the survey, target groups should be identified precisely, offering both demand and supply in sharing economic services, according to the age, level of education, personal attitudes, preferred values by them, etc. Survey should be performed anonymously between the population aged above 15 years. Design of data collection for the quantitative sample survey of the sharing economy, we propose to ensure the following methodological parameters, which will ensure a sufficiently precise description of respondents:

- Number of respondents: 3000 respondents.
- The research should be stratified with the same probability of selection in the four main stratums according to the size of the municipality (less than 1000, 1000–9999, 10,000–99,999, and 100,000 and more).
- Selection method: random selection.
- First of all, a face-to-face interview, where recording respondents' views in the paper questionnaire should be captured.
- Estimated cost of the survey: 1,000,000 CZK without VAT, which means approximately 333,33 € without VAT

In the future, we suggest to use the alternative data collection provided by computer-assisted telephone interviewing or by electronic survey.

6.3. The analysis of regulatory indicators

Using the Conceptual Sharing Economy Model (CSEM) [13], the effectiveness of sharing economy should be researched. It is the basic model that allows to point out the behavior and relationships of economic entities. It is based on the theory of economic subjects and describes the relations of these entities in the environment of sharing the economy. It should also be used to provide recommendations for regulatory measures in the context of public interest theory and enforcement of socially effective behavior (including the limitation of negative externalities). Its basic schema is shown in **Figure 1**. The basis of the model is the largest concentric circle—a true sharing economy. It means that it is an economic model based on the sharing or leasing of products, as opposed to their exclusive and indivisible ownership. The basic interaction in the model is interaction between households and businesses. On the left side of the graph, the impact of households dominates, and in the right part, the impact of the behavior of the enterprise and the business environment prevails.

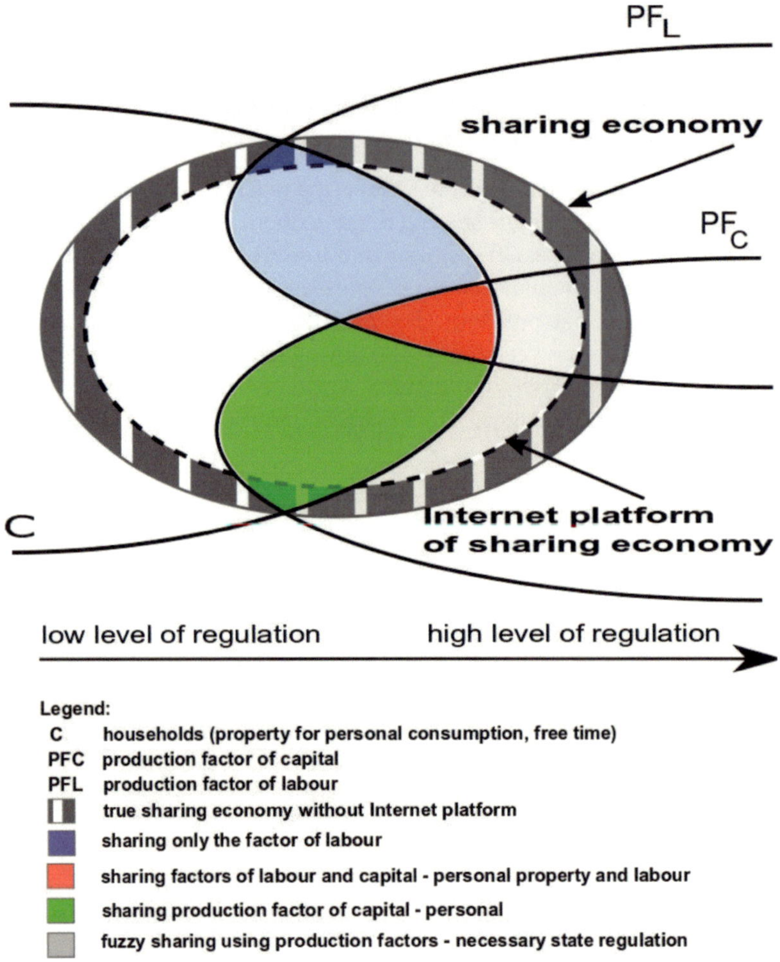

Figure 1. Conceptual Sharing Economy Model—CSEM.

In the description of the model, we can deduce two possible approaches of Internet platform operators (context model) to create support for their delivering. The first approach is to optimize shopping behavior and social efficiency. There are open platforms where information is not hidden for any reason; it is not necessary to share it with other market participants, including state institutions. There are preferred innovation processes, product innovations, as well as helping in a social order to support innovative business activities.

The second approach plays the role as a tool for controlling market activities. From the theory of public interest, there is a significant constraint due to the fact that these platforms produce negative externalities. While, for example, extra charges received in the form of a local tax on accommodation in traditional hotels are returned to the municipal budget, which generally subsidizes local public activities, it contributes to the increase in attractiveness, interest, and it contributes to the increase in attractiveness, interest, and quality of accommodation; in case of sharing, this will not happen. The sharing of accommodation provided by the private sector cannot be taken into account as a driver of progress because it does not contribute to the attractiveness of the service in these ways. It is necessary to assess the need for regulation and choose what is more appropriate—to support traditional business or activities provided by sharing economy. From a control point of view, the most important gray area is the area to the right of the C curve. This "fuzzy sharing" area is a gray zone, an area of social interest in regulation. It is a gray economy that is not socially effective.

6.4. An analysis of the role of science park

The support of spin-off companies also provides science parks. Science parks play a very important role when setting up university and nonuniversity spin-off companies.

The result of such cooperation [30] creates a basis for suggestions formulating, based on best practices (prepared by experts from Taiwan who have been familiar with the situation regarding the already mentioned Bilateral Agreement, for example, the University of Zilina and Broker Center of Air Transport visit), as follows:

- State support, regional, and local autonomy is essential when building science parks, which should be built according to the situation in state development.
- It is important to provide infrastructure and management of high quality while building science parks.
- It is necessary to find qualified experts for effective communication. Experts who will be able to discuss park's business interests suggest examples of "best experience" either from the science park environment or abroad.

This should also be applied to the Broker Center of Air Transport for knowledge and technology transfer with the aim of developing transport and transport infrastructure.

- Autonomy is very important for a science park (in Taiwan, this autonomy is created by companies that are located directly in a science park).
- Ideal equipment should include high-quality, fast Internet connections, free zones, hot links available 24 hours a day, restaurant, coffee shop, shopping area, relaxation center, etc.
- It is advised and effective to learn from the experienced ones when it comes to knowledge transfer and a spin-off company setup. Many people are still expected to "learn from the experienced ones." For service or product providers, it should be a challenge in order to progress and successfully deal with the problems and obstacles we are experiencing.

6.5. Evaluation of the conditions of technology transfer and their commercialization

Next, we also evaluate the possibilities of technology transfer and their commercialization, implemented through spin-off or start-up companies based on the evaluation of experiences from examples of best practices. To assess the current situation in the attitude of SMEs to implement or innovate IT as a basic support of their IS management, we evaluated the opinions of 189 SME managers and of 29 representatives of IT companies (data collection realized by electronic survey and by structured interviews), and data analysis was provided using validation of hypotheses, where we used the statistical methods and validation tools as a part of the SPSS program.

Although many surveys are mostly oriented toward developed and innovative regions, our research activities have been oriented to spin-off companies in economically less developed regions or those with lower innovation activities. While such regions are mostly characterized by the absence of excellent research, the influence of spin-off companies is not only symbolic but instead helps to improve the flow of knowledge within the region. Nevertheless, the number of university-related spin-offs is relatively low, but its interconnection with universities demonstrates that universities can be considered as catalyst for the business sector within the region. Many spin-offs are oriented toward providing consulting services, which in turn could help develop specialized expert services or groups offering knowledge-intensive services within the region. It also supports network creation with other companies and expanding personal contacts between specialists and practitioners. Newly established spin-offs could help develop a second generation of technologically advanced spin-offs in the future.

The worse situation is in the area of other small and medium-sized enterprises (SMEs), which play a significant role in the sharing economy, especially in less developed regions. According to our previous research (realized in 2015 with managers of SMEs), only a small part of SME was interested in the new technologies. We tried to find the key barriers, which prevent companies to utilize the potential of new technologies and innovations. In [31], dealt with a similar issue, they stated that managers of SMEs (representatives of 50 small and medium-sized enterprises located in Lower Silesia) declare lack of financial resources and infrastructure for the comprehensive implementation of integrated information systems to facilitate interpersonal communication in plants. They use only traditional tools, their level of skills is low, and their awareness of staff production facilities in the field of information security, especially when working on the computer and using traditional media, is low, too.

We analyzed respondents' answers to the question if companies have enough information about technologies and innovations. The result is that in two of the abovementioned companies, companies see the low level of satisfaction with the rate of their awareness. Using the Wilcoxon signed rank nonparametric test, we considered a mutual relationship between awareness of new technologies and innovations. As we found, $\alpha = 0.127 > 0.05$, we want to state that they feel better informed about technologies than about innovations. Then, we reviewed the frequency of seeking information about technologies and innovations. We considered the answers to the question "How often do you search for information about technologies or innovations?" We tried to find out if they often seek information daily, weekly, or monthly or prefer to seek information once a year or never. By Wilcoxon signed rank, we tested the frequency of seeking information. We calculated that $\alpha = 0.827 > 0.05$. We found out that they are mostly seeking information about innovation or technologies (approximately monthly, 29%, or half yearly, 20.5%). That is why we can state that if companies are interested in new technologies and innovations, they will be more interested in new information about them.

We have also examined the barriers that restrict the widespread use of IT, system integration of IS, and tools supporting business process automation based on arguments of respondents—company managers and representatives of IT companies. We can state that most of SMEs (93.7%) use basic software support for business doing. This creates conditions that increase the importance of implementing electronic business models also in the small and medium sector as a basic tool for data processing and information search and for their distribution and sharing. This is important when they want to provide their services effectively in strong competition.

Before implementing software for business support, it is necessary to acquire adequate information about the company. We classify this information in the following order: finding functionality to use software support, the level of integration with other tools, identifying areas of corporate activities, and the price they are willing to invest in ICT support, technical requirements, existing IT support, and other requirements included—bottlenecks in business processes, knowledge of work practices, current processes, and planned changes, as well as application to be integrated. If companies have this information, they also create the conditions for the transfer and dissemination of knowledge across the enterprise. In the sharing economy, it is more important, as SMEs are the initiators of changes. It is due to the ability to look for suitable solutions at low cost to the innovation and implementation of electronic ways of connectivity and communication.

7. Conclusion

The digital economy creates not only new opportunities but also new challenges for sharing economy. Companies have developed online platforms (such as Uber and Deliveroo) that other people can use to fill their needs sharing some products or services. These products are then shared between these companies as well as providers. These companies gain an advantage over traditional companies as they benefit from providing their own equipment or service. This reduces costs for them and creates opportunities for providing some "on-demand" services. Their own interest is their technical support and online accessibility.

Evaluating the research findings allows us to define the procedures and conditions that lead to full professional support of sharing. The newest developed technologies are often not fully functional, and related problems with their cooperability need to be solved in practice. Licensed technology addresses these issues and is also a guarantee of their functionality. However, it is well known that companies are not willing to pay for such licenses.

The spin-off companies are the best way to realize the complex development of a functional product that can be later evaluated on the market and offered to a wide range of business partners. The license provides the contractual relationship between the university and the spin-off company, which enables them to further develop and thus contribute to the maintenance and development of the local or regional innovation ecosystem. Managers get the opportunity to understand how to set up and develop an innovative business and what kind of relationships between the university and the spin-off company (especially in the legal and financial field) can arise in transferring knowledge from the university environment into practice. It builds and also strengthens the mutual nonfinancial relationships between the university and the spin-off, where the effect of applying science to business marketing and improving the quality and scope of practical education is done.

According to the OECD Recommendation [32], we can state that a sharing economy is a reality, which makes no sense to distinguish between classical and online activities, but it is necessary to focus on predicting and preparing for digitization, looking for ways to take advantages of digital economy, boosting trust across sectors in a network, complex and global ecosystem. Furthermore, it is necessary to move forward in the digital agenda, in four key policies, as to be opened toward the Internet and innovation, confidence in the digital economy, the construction of a global interconnection, and creating jobs and skills.

Some areas of the sharing economy can impact that they can be positively rated from a certain point of view, from another negative one. It is given by the following aspects:

- Because of offering lower transaction costs and a desire to reach out to the young generation in particular, the prices fall in comparison with payments for traditionally provided services; it can be positively rated by consumers, but it evokes a deflationary impact.
- Platform operators require a lower level of regulation of selected activities, especially services, which may be positively perceived by the service provider, but it creates a negative impact on consumer protection or fiscal interests.

The importance and the need to dispose with regulatory tools and regulatory measures is growing. In this way, the use of the Conceptual Sharing Economy Model (CSEM) plays its role. Managers get a tool for evaluating their own business, where the use of IT offers sophisticated ways of decision-making. They can better decide which regulatory measures or innovations are to be preferred. The result is the ability to get up-to-date information that actually informs about the current business situation.

As a result of our research activities, we recommend the managers to realize:

1. population research aimed at the supply and demand for a sharing economy by sectors, including an estimate of the size of individual markets
2. analysis of trends in individual sectors of the sharing economy (e.g., Airbnb), which would provide an estimation of the size of these segments on an annual basis, which will provide an overview of ongoing service offerings, including occupancy analysis
3. Market sensitivity analysis, including spatial elasticity, based on data obtained from regions, enables partial analyses as a basis for assessing the impacts of the sharing economy on specific entities, including the worsening of their economic results

The statistical evaluation of the practical experience allows for better conditions for the commercialization of technology and also for identifying and making recommendations for the regulatory treatment of the sharing economy.

Sharing economy is not only an alternative distribution and user model but also a competitive alternative to entities offering distribution and use of products, services, time, skills, or competencies. It is due to the fact that some online applications allow feedback and flexible evaluation of the quality of the product or service provided. Based on this information, they may affect the quality of the provider or exclude poor-quality products or service providers from the offer. Even this element of self-regulation could be used to support the limitation of regulatory measures.

Acknowledgements

This chapter contains the outputs of research results supported by following grants: APVV-14-0512 Universities and economic development of regions (UNIREG) and VEGA 1/0087/18 Strengthening the competitiveness of the region of Zilina by improving the impact of University of Zilina in the region and the city of Zilina.

Author details

Mária Pomffyová[1]*, Mária Rostašová[2] and Vladimír Krajčík[3]

*Address all correspondence to: maria.pomffyova@umb.sk

1 Institute of Managerial Systems in Poprad, Matej Bel University, Faculty of Economics, Banska Bystrica, Slovakia

2 University of Žilina, Slovakia

3 College of Entrepreneurship and Law, Prague, Czech Republic

References

[1] Benneworth PS, Charles DR. Overcoming learning uncertainties in the innovation process: The contribution of clustering to firms' innovation performance. In: Oakey R, During W, Kausar S, editors. New Technology Based Firms in the New Millennium. Vol. 3. London: Pergamon; 2004

[2] Dervojeda K et al. The Sharing Economy: Accessibility Based Business Models for Peer-to-peer Markets. European Commision: Business Inovation Observatory Contract No. 190/PP/ENT/CIP/12/C/N03C01 Case study 12. European Commision; 2013. Available from: https://ec.europa.eu/docsroom/documents/13413?locale=sk [Accessed: 2018-05-12]

[3] Eckhardt GM, Bardhi F. The sharing economy is not about sharing at all. Harvard Business Review. 2015;**1**(1-2):12

[4] Bartková L, Veselovská L, Zimermanová K. Possible solutions to dual quality of products in the European Union. In: Proceedings of TRANSCOM; Slovakia, Žilina; 2018

[5] Guarda D. Circular Economy and A Blockchain Driven Sharing Economy? [Internet]. 2016. Available from: https://ipv6.net/news/circular-economy-and-ablockchain-driven-sharing-economy/?shared=email&msg=fail [Accessed: July 12, 2017]

[6] https://falkvinge.net - official site - Rick Falkvinge [Accessed: May 5, 2018]

[7] Botsman R. Defining The Sharing Economy: What Is Collaborative Consumption–And What Is n't? Fast Company. [Internet]. 2017. Available from: https://www.fastcoexist.com/3046119/defining-the-sharing-economy-what-is-collaborative-consumption-and-what-isnt [Accessed: September 5, 2017]

[8] The Shared City, Conference Papers. [Internet]. 2015. Available from: http://resite.cz/download/programs/CZreSITE2015_web.pdf. June 18-19, 2015; Prague

[9] Novotný O, Voříšek J. Digital Road to Prosperity. 1st ed. Prague: Professional Publishing; 2011

[10] Benda M. Sharing economy perspectives. Financial and Economic Information - MF Documentation Bulletin. 2016. MF CR

[11] The UK has Emerged as a Sharing Economy Hub. [Internet]. 2016. Available from: https://www.pwc.co.uk/issues/megatrends/collisions/sharingeconomy/outlook-for-the-sharing-economy-in-the-uk-2016.html [Accessed: May 8, 2018]

[12] PwC Studies: PwC. [Internet]. 2015. Available from: https://www.pwc.com/us/en/technology/publications/assets/pwc-consumer-intelligence-series-the-sharing-economy.pdf [Accessed: June 15, 2018]

[13] Veber J, Krajčík V, Hruška L, et al. The Sharing Economy. Prague: VŠPP; 2016

[14] Peer-to-peer Rental: The Economist. [Internet]. 9.3.2013. Available from: http://www.economist.com/news/leaders/21573104-internet-everything-hire-rise-sharing-economy [Accessed: March 15, 2018]

[15] Čorejová A, Rostašová M, Čorejová T. Knowledge transfer model and spin-off company set up in significant academic centres in Taiwan. In: TRANSCOM 2017: International Scientific Conference on Sustainable, Modern and Safe Transport; 2017; Elsevier Ltd

[16] Rehák Š. Academic entrepreneurship and university spin-off companies. Public Administration and Society. 2009;**10**(2):160-171

[17] McQueen DH, Walmark JT. Spin-off companies from Chalmers University of Technology. Technovation. 1982;**1**(5):305-315

[18] Callan B. Generating spin-offs: Evidence from across the OECD. STI Review, Issue 26. France, Paris: OECD Publishing; 2000. p. 18

[19] Degroof JJ, Roberts EB. Overcoming weak entrepreneurial infrastructures for academic spin-off ventures. The Journal of Technology Transfer. 2004;**29**(3-4):327-352

[20] Klofsten M, Dylan JE. Comparing academy entrepreneurship in Europe – The case of Sweden and Ireland. Small Business Economics. 2000;**14**(4):299-309